Seeing Gravity
The Gravitational Optics of the Universe

" The cover comes from *Harmonia Macrocosmica* by Andreas Celaruius illustrating the Ptolemaic theory of epicycles. That theory addressed observations showing that the orbital progress of the planets was irregular when viewed under the assumption that the earth was the center of the universe. Epicycles are analogous to present day inflation, dark matter, big bounce, and whatever else is required to salvage a hopelessly irreparable standard model.

Preface

> It is only through personal mistakes that new discoveries are made.

It has been five years since the first edition has appeared. In that time, the book made it to third place in the BookAuthority's list of the all time 100 best books on General Relativity with a 4.14 average. It remained their until one of their "authorities" actually read the book, and it was quickly removed from the list. Among their primary collaborators is Elon Musk. That says it all! The book also made it to the Black Holes list, and, again every trace of it was removed when it was realized that the book is anti-general relativity and anti-what conventional wisdom tells us about the existence of black holes.

I thought about a second edition of the book when I realized an error in Chapter 2, Eqn (2.5) in that the Newtonian slope is either a straight line or a circle, depending on how the trajectory is traversed. It, in fact, makes the radius of curvature, (2.1), infinite! It is anyone's guess why this error was never picked up in the almost infinite commentaries on How Newton would have proceeded with his proof of the derivation of the elliptical orbit given his inverse square law for the central force. The error I bear responsibility for is in the transition from Eqn (2.6) to Eqn (2.8): Dots cannot replace primes, i.e., the derivative with respect to the angle variable cannot simply be replaced by the time derivative. Thus, came to be the writing of a second edition.

The first thing was to correct "Newton's slope" $r'/r = \pm \tan \theta$, where θ is the complementary angle that the tangent to the curve makes with the trajectory of the curve. Then the question arises as to why all the proofs of the derivation of the ellipse go through two integrations of the denominator of the radius of curvature when it is set equal to a constant. This constant supposedly incorporates the fact that the central force is inverse-square. Why not just integrate the equation of Newton's slope.

One of the arbitrary integration constants in the derivation of the equation of an ellipse is not so arbitrary for it represents the existence of a central potential. In its absence, the ellipse degenerates to a straight line as it should. Then a comparison with Weber's force brings up the interesting point that whereas Weber's force takes into account the orientation (angular dependency) of two relative charges in motion, it does not take into account their relative rotational energies. Once this is accounted got, Weber's force is seen to consist of the relative contraction of the particles in the direction of motion, the radial acceleration, and the angular momentum, just like in Newton's equation for the force. But, unlike Newton's mechanical equation, Weber's equation introduces a limiting speed–more than half a century before relativity was born.

Generalizations to Newton's slope arise when the metric coefficient of the angular term in the metric is no longer proportional to the area of a circle. A major distinction between elliptic and hyperbolic planes of constant curvature lies in the fact that the latter can support both open and closed trajectories. Considering the latter, we find the equation of a soliton, or the equation of a Joukowski ellipse. Apart from the technical area of the aerodynamics of wing foils, the Joukowski transform rose to prominence in the transformation of a circle to an ellipse in the complex plane. The resulting ellipse has its center at the origin and has a Hookean central force. The square of the Joukowski ellipse is another ellipse that displaces the origin to the focus of the ellipse and with it the transition from the Hookean law to a Newtonian inverse square.

We will discuss these dual laws henceforth. But in the new chapter on curvature, we obtain a non-traditional form of a conic independent of any constant of integration. The nonlinearity of the hyperbolic plane has brought in particle like solutions called solitons that propagate undistortedly on the surface of constant negative curvature called a pseudosphere. The Binet equation implicates a central force of inverse-fifth, which is the only self-dual law that exists at small radial distances, while distortions occur at greater distances.

We use Newton's impact method to derive the force laws both in the elliptic and hyperbolic planes of constant curvature. There is no simple prescription that allows us to replace the radial distance in the Euclidean plane with the Lobachevskian "straight" line in the hyperbolic plane.

Another property will we investigate is the appearance of cusps in the non-Euclidean planes of constant curvature. Cusps in otherwise continuous trajectories first appeared in epicycles where a close trajectory orbits a larger closed trajectory. Cusp also appear in Euclidean trajectories, notably the cardioid which has a heart-shaped trajectory. When the force is attractive, the Binet equation has the form of Einstein's modification for the deflection of light. However, we would expect such a trajectory to be open, and not closed, and this will will discuss in due course.

If we allow for nonlinearities in the metric coefficient involving translation, we get generalized Ampère's laws and surfaces of revolution. Whereas Weber's force does not lead to a hyperbolic surface of constant curvature, its generalization does. Again the Joukowski ellipse appears in conjunction with the pseudosphere, and the possibility of particle like-solutions called solitons. Particle solutions always appear when there is more than one stationary solution since no stationary can be globally stable.

In fact, gravitational force have long been noted to be associated with the pseudosphere. It has been known since Newton's time that a doubly connected bugle can defy gravity when it ascends a ramp at large inclination angles. Also these amazing properties were documented as early as 1694, the above list of phenomena are new additions, some of which are still not clearly understood, that we will touch upon in this second edition of *Seeing Gravity*.

Pervolia, August 2024   Bernard Lavenda

Preface to the first edition

The first question that might come to mind when the reader picks up a copy of this book is: Why write a book on the optical properties of gravity? Optics is what you see, whereas gravity is what you feel. Newton kept them distinct, dedicating *Principia* to the laws of physics and gravitation, and *Opticks* dealing with a "Treatise on Reflexions, Inflexions, and Colours of Light."

Curiously enough, Einstein's first attempt to explain the deflection of light by the Sun employed Snell's law (and not Huygens' principle as he assumed) where he claimed that

> The principle of the constancy of the velocity of light holds good according to this theory in a different form from that which usually underlies the ordinary theory of relativity.

This would mean that a space varying gravitational field will decrease the speed of light in its neighborhood just like a varying index of refraction. Whereas in his subsequent formulation of the General theory of Relativity (GR) he assumed, following Poincarè, that gravity like light travels at the speed of light. And like light which is propagated as electromagnetic waves (EM), gravity should also be propagated as waves, Gravitational Waves (GWs).

Specifying that all particles follow geodesics made it equivalent to geometrical optics, which saw him derive a modified equation of a Keplerian orbit, in which he obtained numerical correspondence both with perihelion advance of Mercury and the deflection of light by the Sun in one fell swoop.

This seems rather incredible that the same law should hold good for a heavenly body as for a ray of light, by only neglecting a single term in that equation. In fact, it is the optical analogy to gravitation that is the thread that binds Newton's theory of gravitation with Einstein's General Relativity. Moreover, there is no reason to stop at Einstein's modification of the equation of a Keplerian orbit, thus opening up a whole range of possibilities from Kepler's conics to Cassini ovals, and beyond.

A new branch of optics, dealing with dielectric materials, known as metamaterials, has benefited greatly from the analogy with relativistic mechanics. New phenomena, such as cloaking and perfect imaging, are consequences of the fact that when light rays are focused by a lens they do not conform to Euclid's fifth postulate, but, rather, display a vast range of non-Euclidean behavior: Light perceives a medium as a curved space, and, so too, gravity perceives a medium of curved spacetime.

In all fairness, the idea is not new but can be traced back to, of all people, Maxwell who, in 1854 gave an example of an absolute instrument using spherical geometry.

Although Hooke and Newton were personal adversaries, Hooke's linear law and Newton's inverse square law are duals to one another, in the exact same way that a Luneburg lens is the dual of an Eaton lens. Moreover, Luneburg showed that ellipses in the plane were stereographic projections of a perfect optical instrument on a sphere, which is none other than Maxwell's example of an absolute instrument. So already optics has been given generalizations of Keplerian orbits when we 'lift our eyes to the heavens,' where light can travel in circular orbits. Is there a corresponding analogue for gravitation? Is the theory limited to conical sections, or are there a generalizations of orbits obtained from toridal sections, since a torus is also part of a set of Riemann surfaces?

Like optical and gravitational analogies, there must be electromagnetic and gravitational analogues based on the very nature of light. When Ampère came on the scene at the dawn of the nineteenth century people were concerned about the otherwise instantaneous propagation of the Coulomb force, which was resolved by Weber in his law of force of moving and static charges. Although many astronomers tried to apply the same equation to gravitation there was no common consensus, and Laplace's argument that gravity propagates some 7 million times faster than the speed of light could hardly be refuted. Any finite speed of propagation of the gravitational force would mean that it would display diffraction phenomena, and, in particular, aberration that would deflect the action of an otherwise radial force.

An illustration of the union of the EM and GW was found by a German school teacher named Paul Gerber just before the turn of the twentieth century, who was able to get the correct advance of the perihelion of Mercury. However, no one realized that his

modification of the equation of the orbit was precisely the Weber force of electrodynamics, and it moreover predicted that gravity travels at $1/\sqrt{3}$ the speed of light. This
was later confirmed by Schroedinger, over a quarter of a century later, in his attempt
to by-pass GR using only classical mechanics. Whereas Schroedinger knew the value
of the parameter in advance, Gerber's matching with Weber's force would have constrained it to be that value, and none other. This would no longer have been considered
'gap' fitting.

After developing these analogies it will bring me to the comparison of EM waves and
GWs. Both supposedly propagate at the same finite speed, yet the former can propagate in the vacuum while the latter is viewed as 'ripples' of space-time, Einstein's
new ether. EM waves can be shielded; GWs cannot. EM waves manifest diffraction
phenomena; none is known for GWs. The energy of EM waves can be localized; GWs
cannot. Both are considered transverse waves with two states of polarization.

And here is where I can offer some novelty and reserve. I give what can be considered
a one line derivation of the Peter-Mathews expression for the decrease of the period of
a binary that they derived from a laborious procedure in GR, and show the luminosity
to be four orders of magnitude higher in aberration than that of black-body radiation.
That translates into an unheard of Stefan law of T^7, compared to black body which has
a T^4 law where T is the absolute temperature. And it predicts that GWs have the same
number of degrees-of-freedom as GR without having to introduce pseudo-tensors, or
worry how a Poynting vector can be employed when gravitational energy cannot even
be localized. All this would tend to imply that GWs are much closer to their EM wave
cousins than would have been expected.

In the journey I take in this book I have been influenced by relatively few people
with the notable exceptions of the late Tom Van Flandern, whose physical intuition I
concur with, Angelo Loinger, whose insistence on the "purity" and limitations imposed
by the founding fathers of GR should not be messed with, and to my late friend, Fred
Cooperstock, who believed that "Electromagnetic waves have an intrinsic duality; they
are necessarily also gravitational waves."

This book was (and maybe still is) under contract for publication with World Scientific Publishing Company. After having submitted the manuscript, the in-house editor

decided to send it out for review–a procedure that should have been carried out *prior* to the stipulation of the contract. After several months of delay, five reviewer comments came back without a single specific criticism–just a common feeling of 'concern.' The concern was over my criticism of the 'discovery' of gravitational waves, which prompted the in-house editor to 'suggest a major revision' wherein all 'unwarranted' criticisms of general relativity should be removed. This was nothing less than a 'gagging' order–something unbecoming of a publisher who claims to be 'neutral.' The fear of readership decline is enough to suppress any and all criticisms of a theory that has been pushed beyond its limits. The publisher should have thought of that prior to the publication of my other two books: *A New Perspective on Relativity: An Odyssey in Non-Euclidean Geometries* and *Where Physics Went Wrong*, which are along the same lines. It is fairly safe to say that Einstein, would he return today, would not recognize his own theory, and, moreover, would be appalled by some of the results that its numerical 'extension' has obtained that contradict the basic premises of his theory.

Pervolia & Shoresh, April 2019   Bernard Lavenda

Contents

1 Introduction

1.1 The present state of the art

A half a century ago, General Relativity (GR) was a side-show of a side-show. Even Einstein didn't think of his theory as anything more than rounding off Newtonian theory in being able to predict minute effects like the advance of the perihelion of Mercury. In a somewhat apologetic mood, Einstein wrote in his Forward to Bergmann's book: [1]

> It is true that the theory of relativity, particularly the general theory, has played a rather modest role in the correlation of empirical facts so far. . .
> It is quite possible, however, that some of the results of the general theory of relativity, such as the general covariance of the laws of nature and their nonlinearity, may help. . .

How the times have changed!

What Einstein was referring to was quantum theory which hogged center stage for the remainder of Einstein's life. Controversy still surrounded the existence of GWs In the 1957 Chapel Hill conference, Feynman was successful in convincing the majority of participants that GWs do, in fact, exist. He did so by using a 'sticky bead' analogy that assumed *a priori* that GWs carry energy which could be transferred to the sticky beads that would show up in frictional heat. In those days, physics was done by voting.

Yet, the energy stress tensor in Einstein's field equations contains all forms of energy–except gravitational energy. It turns out that this energy is supposed to be accounted for by a gravitational pseudo-tenors, which being composed of the metric and its first derivatives does not contain the necessary second derivatives that would make it a tensor.

Pseudo-tensors, like the Christoffel connection coefficients of which they are composed, can be made to vanish by a mere change of coordinates. They can even appear

---

[1]P G Bergmann, *Introduction to the Theory of Relativity* (Prentice-Hall, Englewood Cliffs NJ, 1942)

in Euclidean flat space through a choice of coordinates, for example, cylindrical coordinates. So even if one is willing to accept the existence of GWs, they must be very different than how GR realizes them.

GWs are ripples in spacetime, and unlike electromagnetic (EM) waves, they need a medium to propagate in, yet they travel at the same speed as light. But, maybe the GWs are not the same as the gravitational force, and whereas the former propagate at the speed of light, the latter propagates instantaneously.

This would be analogous to the Coulomb field acting instantaneously while the EM fields propagate at the speed of light. But when Coulomb proposed his law in strict analogy with Newton's inverse-square law, the connection between electricity and light was unknown.

In Weber's (Wilhelm not Joseph) theory of EM both static and motional fields found coexistence through the introduction of a constant as a limiting speed of propagation of electric charges. The same approach appears in the Lorentz force which considers the sum of a static Coulomb potential and a motional magnetic field. Allowing the constant to tend to infinity would reproduce Coulomb's law.

Is there an analogous expression between Newton's law, and what would be the gravitational analogue of the Grassmann force? Although this would support the nature of the supposed polarization of GWs, it would introduce a lot of other problems because GWs wouldn't manifest the same optical characteristics as light, once beyond the geometric optical limit of small wavelength. It would necessary implicate an orthogonal field to the gravitational field, which has been called a gravito-magnetic force in analogy with the magnetic field.

The explanation given by LIGO of what causes GWs, like the collision of binary black holes, is based on Numerical Relativity (NR) which is a nascent field that uses computers to solve Einstein's equations numerically from prescribed initial data. For example their exist numerical codes that suggest a collapsing star emits between 1 and 2 % of its total mass in the form of GWs.

Which part of Einsteins equations contain energy dissipation and radiation is left unspecified. All this is very surprising in view of the fact that GR can't even solve the two-body problem. And what is even more remarkable that it confirms that two colliding impulsive GWs produce a singularity.

The procedure [2] is to discretize the Einstein field equations where arbitrary source functions are introduced that supposedly "encode the gauge freedom of the solution." Black holes are already built into the system by using a "scalar field gravitational collapse to prepare the initial data." So the results they hope to establish are already incorporated into the problem rather than being the outcome of the nonlinear evolution of the Einstein field equations.

It would appear more advantageous and convincing to cast the sourceless wave equation of GWs as an eikonal equation [3] and study the propagation of the wave fronts. But because a linear wave equation has been used, there is no hope in the propagating wave surfaces to develop discontinuities since the nonlinearities have been discarded. It seems hardly realistic to linearize the Ricci tensor of an indefinite metric which incorporates the hyperbolic nature of the resulting waves, and then to discard products of the Christoffel systems in that expression by arguing that the characteristics depend "only on the highest occurring order of derivatives."

1.2 A brief critique of GR

Fock also found exception with the name the General Theory of Relativity, which he considered to be misleading. [4] For, he argued, it is neither 'general' nor does it contain 'relativity.' In GR, "space appears nonuniform whereas relativity relates to uniformity."

Roughly speaking, Einstein's field equations consist in a marriage of the gravitational field, embodied by a modified Ricci tensor in order that its four-divergence vanish, and energy-stress tensor, whose four-divergence would represent the conservation laws of energy and momentum. But, as Einstein was to later admit, the geometrical side of his equation was a marble edifice, while the energy-stress side was a mere wooden shack

---

[2]F Pretorius, "Evolution of binary black hole spacetimes," arXiv:gr-qc/050714.

[3]V Fock *The Theory of Space Time and Gravitation* (Pergamon, Oxford, 1959) Appendix C.

[4]V Fock, *op. cit.* p 350.

in comparison. For it limited the applications to 'dust' or 'perfect fluids.'

Einstein's criterion of emptiness was the vanishing of the Ricci tensor, since the modification was no longer necessary. It was this condition that permitted Schwarzschild in 1916 to find a solution to his equation, but one with mass that entered as an arbitrary constant of integrating his equations.

Since particles in Einstein's world are limited to follow geodesic paths, which is the next closest thing to straight path of particles that have uniform speed in Euclidean space, what his equations describe is a 'dust cloud' of non-interacting, point particles. Obviously, the conservation laws follow trivially. Energy and momenta can be combined into a four-vector, whose time rate of change vanishes. All this is fine and well, but where then does gravitation enter.

Here things start to get messy. It is assumed that "the world line of a particle not acted upon by any forces, except gravitational, is a time-like geodesic." [5] This supposedly replaces Newton's first law concerning a particle's uniform motion when not acted upon by forces.

But isn't gravitational attraction a force? Not in Einstein's theory, 'empty' space is given by Einstein's criterion that the Ricci tensor vanish, which means that "there is no matter present and no physical fields except the gravitational field. The gravitational field does not disturb the emptyness [sic]." [6]

This explains why there is no account of gravitational energy in Einstein's energy-mass tensor, but it does not explain how gravitational interactions are accounted for. In terms of the geodesic equations, the vanishing of the usual three-vector acceleration is relegated to the vanishing of a four-acceleration because the so-called connection coefficients, which contain the metric and its first derivatives, do not vanish.

This interaction, however, [7]

---

[5] P A M Dirac, *General Theory of Relativity* (Wiley, New York, 1975) p 15.
[6] Dirac, *op. cit.* p 25.
[7] Fock, *op. cit.* p 326.

> do not enter explicitly into the energy tensor but is accounted for indirectly through the gravitational potentials. The presence of a gravitational field and of the energy related to it reveals itself, as we know, in a change of the properties of space time. Gravitational energy can be separated out in the form of additional terms in the energy tensor only in an artificial manner by fixing the coordinate system and reformulating the problem in such a way that the gravitational field is taken to be superimposed on a space-time fixed properties, just as is done in Newtonian theory.

The first observation is the 'special' coordinate system which violates one of the basic tenants of GR– covariance– or all physical laws should be independent of the coordinate system used to express them. The second observation is what is the need of 'superimposing' the gravitational field on space-time when it is already built into it through the metric tensor? Third, if the same procedure is followed in Newtonian theory what are the advantages of GR?

Continuing we read:

> The additional terms in the energy tensor that correspond to gravitational energy do not possess the property of covariance (i.e. they do not form a tensor). According to the choice of coordinate system the values of these terms at a given space-time point may prove to be zero or non-zero, which would be impossible for a tensor. Therefore, gravitational energy cannot be localized.

Everything that Fock has said so far is sacrosant. But then he offers as a reason why gravitational energy cannot be localized precisely because "*a gravitational field cannot be screened.* The only way to avoid the action of the gravitational field is to remove oneself further from the masses that are producing it."

This is in flagrant contradiction to Dirac's statement that even in an 'empty' universe (i.e on devoid of all matter), the gravitational field exists and "does not disturb the emptiness."

At most GR offers us a set of geodesic equations expressing uniform motion in a non-

Euclidean geometry that could never generate GWs. And then there is covariance which allows us to find a coordinate system that reduces the motion to zero, like a co-ordinate system attached to a moving charge so that there will be a vanishing magnetic field.

And finally, there is the problem that has still to be overcome that "even the two-body problem, solved so long ago by Newton, has very little chance of being successfully solved in terms of General Relativity." [8]

If the two-body problem has not been solved in GR, on what basis does LIGO support their claim about coalescing black holes (BHs), or neutron stars other than the mass calculated from Kepler's III law? And certainly the inspiraling and collision of such bodies is well beyond the jurisdiction of GR. It appears they picked the wrong theory of gravitation to describe such violent phenomena.

1.3 Is there a difference between GW and EM waves?

As an alternative, we sympathize with the idea that "electromagnetic waves has an intrinsic duality: they are necessarily gravitational waves." [9] We can further imag-ine that GWs are EM waves that have passed through a gravitational field which has modified their speed of propagation through a medium with a non-unitary index of refraction.

When discussing the deflection of light about the Sun, Eddington writes [10]

> There is an alternative way of viewing this effect on light according to Einstein's theory which is to be preferred. This depends on the fact that the velocity of light in a gravitational field is not a constant but becomes smaller as we approach the sun.

---

[8]T. Levi-Civita, *The n-Body Problem in General Relativity* (Reidel, Dordrecht, 1964) p VIII.

[9]F I Cooperstock & S Tieu, *Einstein 'Relativity: The Ultimate Key to the Cosmos* (Springer,Heidelberg, 2012).

[10]A S Eddington, *Space Time & Gravitation* (Cambridge UP, Cambridge, 1929) p 107.

Roughly speaking, gravity forms the terrain and light passing over it is affected by it just like light passing from one medium to another changes both direction and speed.

This is also reflected in GR: GWs follow the same geodesics as do light waves, both traveling at the speed of light. How then is the transition made from light waves to GWs?

The gravitational properties of light were raised early on. [11] In that article, the gravitational effect of an infinitely thin cylindrical pulse of unpolarized light was studied in terms of possible interactions with a light beam. The result was that if the two are traveling in the same direction this is no interaction, while if the two are traveling in opposite directions there was an attraction.

Strangely enough we find the same thing for GWs: "when moving in the same direction, two gravitational waves do not scatter one another. To exhibit scattering, we need two waves moving in different directions." [12] EM and GW waves are not that all different after all!

The method used by Tolman *et. al.* was to employ the geodesic deviation equation which tells how geodesics will tend to spread out or bunch up in a spatially varying gravitational field. This is supposedly caused by the presence of tidal forces, that will cause trajectories to bend toward or away from each other creating a relative acceleration between them. Physicists relate the geodesic deviation equation to the Riemann curvature tensor to the relative acceleration of neighboring geodesics, while geometrists refer to it as the Jacobi equation which shows that the tendency of spread or bunch up depends on whether the geometry is hyperbolic or elliptic, respectively. It is difficult to imagine that tidal forces are associated with non-Euclidean geometries, even those with constant curvature.

Yet, this is what is argued that geodesic deviation between neighboring geodesics is the result of tidal forces that tend to accelerate neighboring geodesics. [13] It is also difficult to see geodesics accelerating since they describe uniform motion.

---

[11] R C Tolman, R Ehrenfest & B Podolsky, "On the gravitationl field produced by light," textitPhys Rev **37** (1931) 602.

[12] R D'Inverno, *Introducing Einstein's Relativity* (Clarendon Press, Oxford 1992) p 281.

[13] D Tatzel, M Wilkens & R Menzel, "Gravitationl properties of light."

About the same time as Tolman was working on geodesic deviation for light, Levi-Civita [14], and slightly earlier Whittaker, [15] showed that the metric employed in the Einstein field equations, giving rise to the geodesic equations, is not a characteristic surface of a GW but one of EM. Moreover, even if both existed and travel along the same null geodesic it would be extremely difficult to tell them apart.

So how do you tell EM and GW waves apart? Any textbook on GR [16] will tell you that the linearized Einstein's equations have the energy-stress tensor that still represents the source. However, in order to calculate the energy-momentum you must switch over to the gravitational *pseudo-tensor*, that is absent in the field equations! Then, to calculate the energy loss through radiation you have to re-introduce the energy-stress tensor and solve the retarded solution of the linearized wave equation. And after all this, the interaction between GWs and particles is *still* described by geodesic equations. It is apparent that coherency has been sacrificed.

1.4 Does a numerical coincidence prove a theory right?

The classic example where numerical matching of the theoretical and observational results proved the theory right was in Einstein's calculation of the deflection of light about the Sun. It is commonly asserted that the gravitational deflection of light is a true test of GR for without it, you don't get the right answer. Using entirely classical methods, as we mentioned in the *Preface*, gave Einstein half the value of what he was later to find a half a decade later using the full-fledged theory of GR. It was this confirmation by Sobral expedition that rocketed Einstein to world fame.

Also in the *Preface* we have mentioned Einstein's first attempt was a camouflaged application of Snell's law wherein the speed of light is reduced by a factor of $1 - \alpha/r$, with α as the Schwarzschild radius. [17] His result was identical to that done by Soldner

---

[14]T Levi-Civita, "*Rend Acc Lincei* **11** (1930) 3, 113.

[15]E Whittaker, "Note on the law that light-rays are null geodesics of the gravitational field," *Proc Cambridge Phil Soc* **24**/1 (1927) 32.

[16]cf. Ohanian & Ruffini, *op. cit.* Ch 5

[17]A Einstein, "On the influence of gravitation on the propagation of light," *Ann d Phys* **35** (1911.

in 1801 who concluded that "If one substitutes into $\tan \omega$ the acceleration of gravity on the surface of the Sun, and the radius on that body is set equal to unity, one finds $\omega = 0.84$ arcsec", whereas Einstein found 0.83 arcsec. And the story can be continued further back in time with Newton's study of the gravitational bending of paths of corpuscles in *Principia*, and Mitchell's study of the shifting of light by gravity.

There are many other instances where numerical mismatching is used as a means of discrediting rival theories. In many instances, the 'right' answer was known, and it was just a matter of getting there. A case in point is Gerber's theory of the perihelion shift, and the deflection of light. That he got the first right and the second wrong is enough to discredit his entire theory. [18]

GR shares in common with Newtonian theory that the acceleration is independent of the mass undergoing the acceleration, and depends only on the mass causing the acceleration. This allowed the same geodesic equations to be obtained for light as for matter. The only difference was dropping a term in the equation of the orbit, and it was done in a slipshod fashion, by allowing the angular momentum, a constant, to tend to infinity. [19]

In the case of the perihelion precession, there was a discrepancy of 43 arcsec/century from the total, observed, value of 5600. The precession of the equinoxes amount to the largest chunk of 5030, followed by 530 due to the perturbation of the planets. The remaining 43 was filled in by GR, but, note without any explanation of the cause of it. It was chalked up to 'relativity', and was thus considered as 'gap' fitting. [20]

We will obtain the Einstein modification of the orbit (2.78), but that is only the beginning. He, and later Moller, [21] obtain the orbital equation in terms of three roots, two of which represent the perihelion and aphelion of the elliptic orbit. Both assume that the third root does not affect the other two roots. However, "for small values of α, they consider the sum of their inverses equal to the constant, α^{-1}, thus effectively making the third root a function of the other two. By symmetry, the two roots should be a function of the third, which invalidate the original assumption. The effect of the shift

---

[18]K Brown, *Reflections on Relativity* (Lulu.com 2018).

[19]N T Roseveare, *Mercury's Perihelion from Le Verrier to Einstein* (Oxford U P, Oxford, 1982.

[20]A A Vankov, "General relativity problem of Mercury's perihelion advance revisited," arXiv:1008.1811.

[21]C Moller, *The Theory of Relativity* (Oxford, Clarendon, 1953) p 351.

is 3α, the same order of magnitude of the effect they want to predict: 3α times the sum of the inverse of the perihelion and aphelion distances.

So the GR value is still an approximation. Years later in 1942, Einstein presenting Bergmann's book acknowledges a role he had in composing it. We gave an excerpt of it earlier. In the book there is another derivation of the perihelion shift which expands the solution as a Fourier series, and bears the marking of the author of the forward. Obviously, the same result is obtained, but more aesthetically arrived at.

Parenthetically, this is not the first time Einstein makes multiple attempts at deriving a pre-determined result. A case in point was his multiple derivation of $E = mc^2$, the last one being in 1952. The original derivation can be found in Poincaré's *Electricite et Optique* (Gauthier-Villars, 1901) where he calculates the recoil of an artillery canon and comes out with the change in energy is directly proportional to the change in mass.

The point is that the true value of the perihelion shift is still unknown because the "real position of current ephemerides is unknown, and our knowledge about the effect verification in the astronomical observations remain, roughly, at the Le Verrier's [*sic*] level." [22]

It was the French astronomer, Urbain Le Verrier, who calculated in 1859 the perturbation on the orbit of Mercury with other planets, and, in doing so, found an anomaly that could not be explained by Newtonian theory.

It was about this time that Weber completed his synthesis of the Coulomb and Ampère laws in which he introduced a limiting speed for the propagation of electric signals that turned out to be $\sqrt{2}$ greater than that of light. Since Coulomb derived his law in analogy with Newton's inverse square law, it didn't take much imagination for astronomers to attempt to apply Weber's law in a gravitational setting by replacing the masses for the charges. True, the interaction was much smaller, but that was compensated by the astronomical distances separating them.

---

[22]Vanikov, *op. cit.*

Since the advance of the perihelion was known,

$$\kappa \frac{4\pi^2 a^2}{c^2 P^2 (1 - \varepsilon^2)},$$

to within a numerical factor, κ, the goal was to find a potential, which was a function also of speed, that would determine this factor. In the expression for the shift, a is the semi-major axis of the ellipse of eccentricity, ε, and P is the period.

The velocity could be introduced through the concept of retarded potentials,

$$V(r, \dot{r}) = -\frac{\mu}{r(1 - \dot{r}/c)},$$

caused by the finite time it takes a signal to be transmitted, where $\mu = GM$ is referred to as the standard gravitational parameter. The mathematical development of retarded potentials had only recently been developed by Lienard and Wiechert so it remained a mere exercise to apply it to the problem at hand. This was accomplished by an unknown German school teacher, Paul Gerber who found $\kappa = 2$, but which gave only $1/3$ the missing value.

However, by giving the potential a 'double dose' of retardation by squaring the factor in the denominator did the trick. The problem then lay with trying to justify the extra 'dosage.' Although there have been numerous attempts to rationalize Gerber's choice, with some fairly recent ones, [23] there still is no rational justification for it.

It seems that Gerber adopted Carl Neumann's idea of a 'flowing' potential, in contrast to what everyone was used to, a 'flowing' momentum. Maxwell in his *Treatise* summed up his perplexity by stating there "is the greatest possible difference between the transmission of a potential, according to Neumann, and the propagation of light." Although Gerber's paper of 1902 discusses at length whether a retarded force or a retarded potential is the correct choice, he remained ambivalent about it.

Notwithstanding the criticisms, we shall show that Gerber's retarded potential provides an external force in the orbital equation that is exactly equal to Weber's force provided the propagation speed of gravity is $c/\sqrt{3}$.

---

[23]N T Roseveare, *op. cit.*

Although Weber's force depends both on the relative velocities of the charges and their relative acceleration, it cannot be considered as a retarded force that would lead to aberrational effects. We will come across one such force in Laplace's calculation of the speed of light where the force is a linear function of ratio of the velocity and the propagation speed of gravity. In contrast, Weber's force is *quadratic* in that ratio, and so it is a radial force. As a matter of fact, it will be the relative acceleration in Weber's force that will determine the secular precession.

Failing to recognize this, Roseveare criticizes Gerber's derivation saying that it doesn't include "velocity dependence of the mass implicit in any viable theory," and when this is taken into account, 7 arcsec per century is added to the perihelion shift thereby invalidating his theory.

Not only does GR not explain the 43 arcsec per century, other than saying it is relativistic, are we to throw out an approach that does not achieve numerical equivalence when extra baggage is added on? Any theory could suffer a similar fate. Moreover, when Gerber's theory was applied to the deflection of light it gave a value twice the observed one adding more fuel to the fire for abandoning it. As Brown [24] puts it, "the predictions of Gerber's potential differ significantly from those of general relativity for other phenomena, such as light deflection which falsifies Gerber's model."

A more recent numerical matching accomplishment is the decrease in the orbital period of PSR1913+16, where it is claimed that "the measured rate of change of the orbital period agrees with that expected from the emission of gravitational radiation according to general relativity to within about 0.2 percent." [25]

The expression the authors use for the rate of change of the period is the one derived by Peters & Mathews [26] from Einstein's quadrupole formula that he obtained considering a spinning dumbbell. We will derive that formula very simply as 5th-order aberration, and show that the luminosity corresponds to a Stefan's law of T^7, in comparison with black body radiation which is a T^4 law.

---

[24]K Brown, *Reflections on Relativity* (Lulu.com 2018) §Gerber's gravity.

[25]J M Weisberg & J H Taylor, "Relativistic binary pulsar B1913+16: Thirty years of observations and analysis," in *Binary Radio Pulsars, ASP Conference Series* **TDB** (2004)

[26]P C Peters & J Mathews, "Gravitational radiation from point masses in a Keplerian orbit," *Phys Rev* **131** (1963) 435-440.

A corrected value, taking into account the small correction for the relative acceleration between the solar system and the binary pulsar system, projected onto the line of sight, allows them to conclude that "measured orbital decay is consistent at the $(0.13 \pm 0.21)\%$ level with general relativistic predictions for the emission of gravitational radiation."

Consequently, if Weisberg & Taylor are correct in the prediction, based solely on numerical matching, it proves that GWs suffer aberration, not like light waves, at 3rd-order, but at 5th-order. And if GWs are the transmitters of the gravitational force, then aberration would cause orbits to gain angular momentum, not lose it through radiation. [27]

Therefore, one must exercise extreme caution in using numerical correspondence to draw conclusions on the validity of a theory. For applying this criterion to planetary orbits there would be no reason to throw out Ptolemy's theory of cycles and epicycles! Cellarius's demonstration of the epicycles is depicted on the cover.

We will now begin at the beginning and trace the optical roots of Newton's theory all the way to GR, and beyond.

---

[27]T Van Flandern, "The speed of gravity: What the experiments say."

2 Curvature: The Geometry of Force

2.1 The Radius of Curvature

When Newton came on to the world stage during the second half of the seventeenth century, the table had been prepared for him by Kepler during the first half. Specifically, Kepler laid down his three laws governing celestial motion, and the time was ripe to determine the force, or forces, involved in guiding the planets in their orbits about the Sun.

It was common in those days to divide the problem into two: The 'direct' and 'indirect' problems. In the former, the path was given together with a center of force, and the aim was to determine the force necessary to maintain that orbit. This was a natural for Newton for he already knew that the orbit was elliptical.

In the indirect problem we are given the force and its center of action, and seek to determine the orbit. This is the problem of actual interest because we have universally embraced Newton's law of gravity as being an inverse square law.

To arrive at this conclusion, Newton concerned himself with curvature: Curvature should be an imprint of force. In the early 1670s, Newton derived an expression for the radius of curvature, ϱ, in polar coordinates (r, ϑ):

$$\varrho = \frac{r(1 + z^2)^{3/2}}{1 + z^2 - z'},\tag{2.1}$$

where $z = (1/r)r'$ is the slope of the curve, and the prime stands for differentiation with respect to ϑ. We will now see how the denominator in (2.1) determines the law of force.

In order to do so, Newton used two additional relations requiring only that the force be directed toward its source. This limits the motion to a plane. Figure (2.1) appeared in Proposition 6, Theorem 5 of the revised edition of the *Principia*.

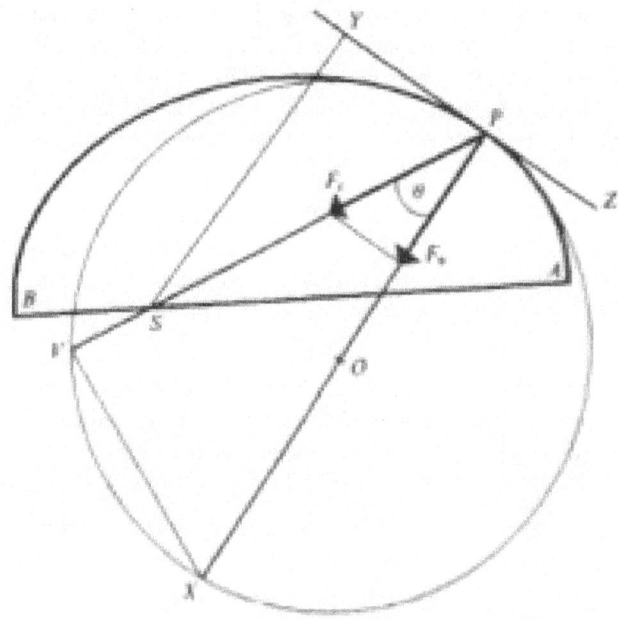

Figure 2.1: Newton's revised diagram for his fundamental theorem for the second and third editions of the *Principia* with the osculatinq circle inserted. Taken from J Bruce Brackenridge, "The critical role of curvature in Newton's developing dynamics," in *The Investigation of Difficult Things* P M Harman, A E Shapiro, eds (Cambridge U P, Cambridge, 1992) p 231-260.

The trajectory APB has a tangent ZPY which is normal to the line PS, where S is the center of force. PVX is the circle of curvature whose radius ϱ is OP.

Newton's Proposition 1 contains Kepler's area law,

$$L = vr \sin \alpha, \tag{2.2}$$

where the instantaneous radius is SP, and v is the tangential velocity at the point P. All central forces conserve the angular momentum L, which is proportional to the area swept out by the satellite per unit time. The angle α is formed from the tangent ZPY and the radius SP. It's complementary angle ϑ is shown in the figure.

The second relation used by Newton is the force that would result if the actual trajectory were to be replaced by the osculating circle of curvature whose radius is ϱ. The centripetal force, F_0 is related to the central force F_c according to

$$F_c = F_0 \cos \vartheta = F_0 \sin \alpha = \frac{v^2}{\varrho}, \tag{2.3}$$

in accordance with Proposition 4. Implied throughout is Newton's assumption of *uniform circular motion*, i.e., constant angular speed as well as centripetal acceleration.

Introducing Kepler's areal law (2.2) into the expression for centripetal acceleration, (2.4), leads to the expression

$$F_c = \frac{L^2}{r^2 \varrho \sin^3 \alpha} \tag{2.4}$$

for the central force. Newton's proof would be complete if he could show that

$$\varrho \sin^3 \alpha = \text{const.}$$

In terms of the complementary angle, α, the slope of the trajectory is

$$z = \cot \alpha \tag{2.5}$$

so that

$$1 + z^2 = 1/\sin^2 \alpha,$$

which appears in the numerator of the expression for the radius of curvature, (2.1). Introducing that expression into the central force results in

$$F_c = \frac{L^2}{r^3} \left(1 + 2\frac{r'^2}{r^2} - \frac{r''}{r} \right)$$

$$= \frac{L^2}{r^3} \left(1 + z^2 - \frac{r}{2}\frac{d}{dr}(1 + z^2) \right). \tag{2.6}$$

In terms of the angle θ Newton's slope (2.5) is

$$z = \frac{1}{r}\frac{dr}{d\theta} = \frac{r'}{r} = \tan\theta. \tag{2.7}$$

Introducing the inverse radial coordinate, $u = 1/r$(2.6 transforms into the Binet equation

$$u'' + u = \frac{F_c}{L^2 u^2} \tag{2.8}$$

In order that the right-hand side be a constant, the central force must be inverse-square, $\propto 1/r^2$.

Rather than dealing with the Binet equation directly (2.8) authors like J Bruce Brackenridge deal with the curvature equation

$$\frac{d}{dr}(1 + z^2) - 2(1 + z^2)/r = -\frac{2}{L^2}(F_c r^2). \tag{2.9}$$

Since the Binet equation establishes that the central force is inverse-square, we can set the right-hand side equal to a constant, $2A$; then letting $f = 1 + z^2$, (2.9) becomes the ordinary inhomogeneous differential equation

$$\frac{df}{dr} - 2f/r = -2A.$$

The complementary solution, obtained by setting the right-hand side equal to zero is $f_c = Cr^2$, where C is an arbitrary constant of integration. Since the particular solution is $f_p = 2Ar$, the complete solution is the sum of the two, viz.,

$$f = f_c + f_p = 2Ar + Cr^2.$$

Reintroducing $1 + z^2$ for f, and writing $B^2 - A^2$ for C, where B is another in arbitrary constant of integration, allows Newton's slope to be written as

$$z = \frac{r'}{r} = r\sqrt{B^2 - \left(\frac{1}{r} - A^2\right)}. \tag{2.10}$$

Integration leads to

$$\theta = \int \frac{dr}{r^2\sqrt{B^2 - (1/r - A)^2}} = \cos^{-1}\frac{1/r - A}{B} - \alpha,$$

where α is another arbitrary constant of integration. Rearranging, we come out with the equation of an ellipse,

$$\frac{1}{r} = A + B\cos(\theta + \alpha). \tag{2.11}$$

From this Brackenridge [1] concludes

> Thus, as Newton has claimed, given the curvature from the force, the path is uniquely determined. Whether Newton could have produced a version of this proof, as Whiteside claims, is a matter of personal conviction.I, for one, have no doubt that he could. But he need not have done so, for the outline provided in Corollary 1 of Proposition 13 is adequate.

Thank heavens that Newton didn't present such a proof! This is nothing in Proposition 13 Corollary 1 regarding his slope (2.7). The only place mentioned is in conjunction of the logarithmic spiral where it is constant.

For the question immediately arises why perform two integrations, introducing two arbitrary constants when one would suffice by simply integrating Newton's slope, (2.7)? Actually, there should be a $\pm$ in that equation to take into account the ambiguity in how the path is traversed. Choosing the positive sign, a simple integration equation results in

$$\ln r = -\ln\cos\theta - \ln B, \tag{2.12}$$

---

[1] J B Brackenrdige, "The critical role of curvature in Newton's developing dynamics," in *The investigation of difficult things*, eds P. M. Harman & A. C. Shapiro, Cambridge U. P., 1992, p. 231.

where B is an arbitrary constant of integration. Equation (2.12) is an equation of a straight line! It is what you get when you set $A = 0$ in (2.11)! The negative sign gives a circle passing through the origin. And we know that the force responsible for that is an inverse-fifth power of the radial coordinate [2]

The Newtonian force, (2.6),

$$F_c = r\dot{\theta}^2 - \ddot{r},\qquad (2.13)$$

where the dot indicates differentiation with respect to time is not comparable to Weber's law between charges e and e'

$$F_W = \frac{ee'}{r^2}\left\{1 - \left(\frac{\dot{r}}{c}\right)^2 + 2\frac{r\ddot{r}}{c^2}\right\}.\qquad (2.14)$$

Weber's force (2.14) takes into account the relative motion of two charges and their orientation but not their angular dependence, and their limiting speed, c. The angular dependence can be introduced through $\dot{r}^2 \to \dot{r}^2 + r^2\dot{\theta}^2$ in which case (2.14) becomes

$$F_W = \frac{ee'}{c^2 r}\left\{\frac{c^2 - \dot{r}^2}{r} - r\dot{\theta}^2 + 2r\ddot{r}\right\}.\qquad (2.15)$$

Whereas the first term in 2.15) is related to the contraction in the direction of the motion, the last two terms are analogous to Newton's law, accounting for radial acceleration and angular momentum. Newton's law is purely mechanical, and there is nothing that would indicate a limiting speed.

In regard to Weber's law, (2.14), the terms involving Ampère's law contain c^{-2}. Ampère established that the ratio between parallel current elements to the force between longitudinal current elements were in the ratio $1 : 2$.

In Weber's *Sixth Memoir* published in 1871, there appears a critical length associated with the reversal of the Coulomb force. In his words: [3]

> when particles e and e' are of the same kind, *they do not always repel*

---

[2]J.M.A. Danby, *Fundamentals of Celestial Mechanics*, Willmann-Bell Inc., 2nd ed., 1988, Sec 4.9
[3]L Hecht, "The significance of the 1845 Gauss-Weber correspondence," *21st Century* (1996) 22-43.

each other; thus when $\dot{r}^2 < c^2 + 2r\ddot{r}$, they repel so long as

$$r > \frac{2ee'}{mc^2},$$

and, on the contrary, they attract when

the inequality is reversed. The mass, m, appears if, for no other reason, than the sake of dimensions. The right-hand side will be recognized as twice the classical electron radius. This was no small feat taking into account that the year was 1871.

In fact, John Michell, a letter to his friend Henry Cavendish in 1783, predicted that once a star exceeded that of the Sun in proportion $500:1$, he prophesied that

> supposing light to be attracted by the same force in proportion to its *vis inertiae*, with other bodies, all light emitted from such a body would be made to return towards it by its own proper gravity This assumes that light is influenced by gravity in the same way as massive objects.

2.2 From Euclidean to non-Euclidean planes

Gauss assumed that any two dimensional surface in a three-dimensional world had, at least locally, a geodesic polar parameterization, (r, θ). Euclidean straight lines are also geodesics with a flat metric

$$ds^2 = dr^2 + r^2 d\theta^2, \tag{2.16}$$

which gives rise to Newton's slope, (2.7). Our interest, however, is closed trajectories, of which the ellipse (2.11) is an example. In order to go beyond the Euclidean plane, we might try hyperbolic and elliptic planes of constant curvature whose respective metrics are given by

$$ds^2 = dr^2 + R^2 \sinh^2\left(\frac{r}{R}\right) d\theta^2, \tag{2.17}$$

$$ds^2 = dr^2 + R^2 \sin^2\left(\frac{r}{R}\right) d\theta^2, \tag{2.18}$$

21

where R is a characteristic scale factor in non-Euclidean geometries.

Unlike the elliptic plane, where all trajectories are closed, the hyperbolic plane admits both closed and open trajectories. According to (2.17), Newton's slope will be given by

$$\frac{1}{R\sinh(r/R)}\frac{dr}{d\theta} = \pm\tan\theta. \tag{2.19}$$

Choosing the minus sign, and integrating gives

$$\tanh\left(\frac{r}{2R}\right) = \coth\left(\frac{r}{R}\right) - \operatorname{csch}\left(\frac{r}{R}\right) = \cos\theta, \tag{2.20}$$

where, for simplicity, we have suppressed the arbitrary constant of integration. Solving for r results in

$$r = \pm 2R\tanh^{-1}(\cos\theta) \tag{2.21}$$

which are two Joukowski ellipses shown in (2.2).

Taking the derivative of (2.21), we find $r' = 2\csc\theta$, and whose integral is

$$\theta = 2\tan^{-1}\left(\exp\left(\frac{r}{2R}\right)\right) \tag{2.22}$$

where R is a constant of integration. Equation (2.22) is just another way of writing the Joukowski ellipse.

From the metric in the hyperbolic plane of constant negative curvature where $G = \sinh^2(r)$, we obtain the equation for Newton's slope as [cf. Eqn (2.19):

$$\frac{1}{\sinh(r)}\frac{dr}{d\theta} = \pm\tan\theta.$$

Now choosing the positive sign, we have the integral:

$$r = -2\tanh^{-1}(\cos\theta) = \ln\left(\frac{1-\cos\theta}{1+\cos\theta}\right).$$

This agrees with (2.22) when we write it in the form

$$r = \ln\tan\left(\frac{\theta}{2}\right) = \ln\left(\frac{1-\cos\theta}{1+\cos\theta}\right).$$

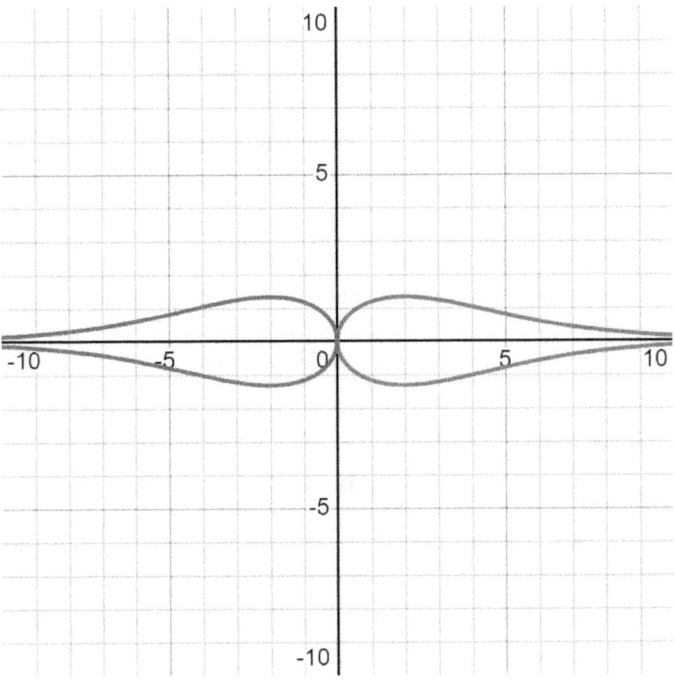

Figure 2.2: Two symmetric Joukowski ellipses in the hyperbolic plane of constant, negative curvature.

In the hyperbolic plane, the inverse of the radial coordinate, $u = \coth(r)$, and $\sqrt{u^2 - 1} = \operatorname{csch}(r)$. Noting that the secant is

$$u + \sqrt{u^2 - 1} = \sec\theta, \tag{2.23}$$

we can add it to (2.20) to get

$$u = \frac{1}{2}(\sec\theta + \cos\theta). \tag{2.24}$$

Equation (2.24) transforms in an analogous way to the one that stretches a circle to an ellipse in the complex plane. In the Euclidean plane, where $u = 1/r$, (2.24) gives an ellipse

$$1/r = \frac{1}{2}(\cos\theta + \sec\theta), \tag{2.25}$$

passing through the origin, as shown in (2.3). Unlike the conventional expression for a conic, containing two arbitrary constants as in (2.11), (2.25) does not contain any constant whatsoever. It is analogous to a circle passing through the origin that is caused by an inverse-fifth force, as we shall now discuss.

The ellipse in the Euclidean plane, (2.25) becomes a Joukowski ellipse,

$$r = \coth^{-1}(A(\cos\theta + \sec\theta)) \tag{2.26}$$

in the hyperbolic plane of constant curvature for the value $A = 1/2$, as shown in (2.4).

Having noticed the similarity between an ellipse through the origin and a circle, we write down the Binet equation

$$u'' + u = \sec^3\theta = (u + \sqrt{u^2 - 1})^3. \tag{2.27}$$

Let us consider the terms in the expression for the secant,

$$\sec\theta = u + \sqrt{u^2 - 1} = \frac{1}{\tanh(r)} + \frac{1}{\sinh(r)}. \tag{2.28}$$

Since the Lobachevskian 'straight' line is $\tanh(r)$, and this gives the Newtonian potential as

$$V = \frac{\mu}{\tanh(r)},$$

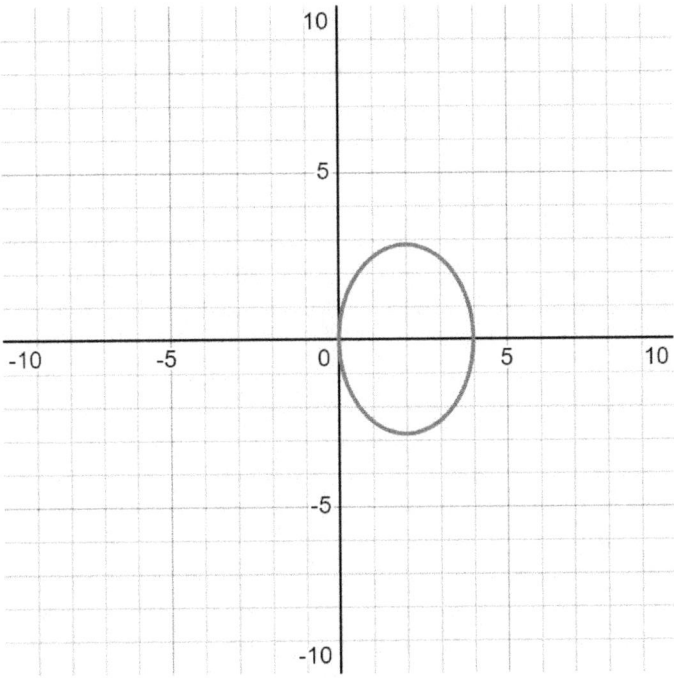

Figure 2.3: The reduction to the Euclidean plane gives an ellipse through the origin.

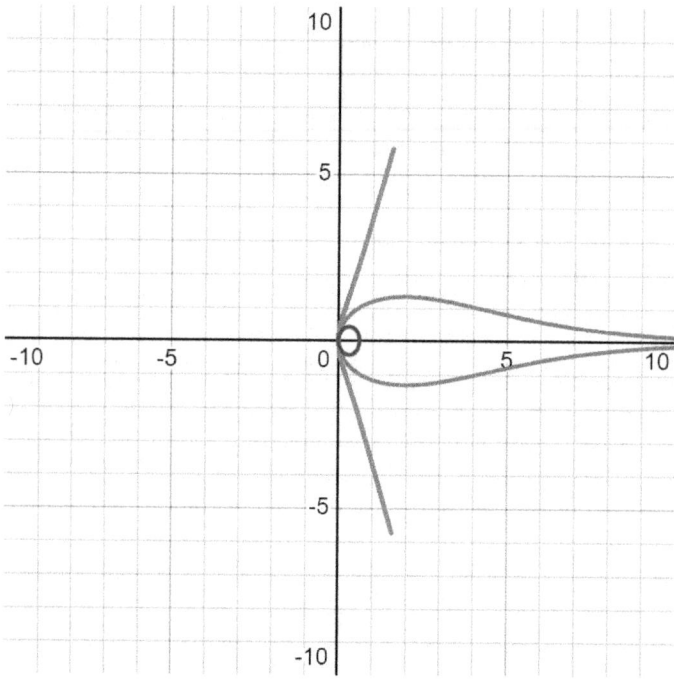

Figure 2.4: The Joukowski ellipse, $A = 1/2$, with its characteristic cusp, characteristic of airfoils, separates open, $A < 1/2$, from closed, $A > 1/2$, trajectories. As $A \to 1$, the ovals transform into ellipses passing through the origin.

where $\mu = GM$ is the gravitational potential. Consequently, the force is

$$F = \frac{dV}{dr} = -\frac{\mu}{\sinh^2(r)}.$$ (2.29)

Thus, for large values of r, the secant will be dominated by the first factor in (2.28), and the force will be essentially and inverse-fifth power of the radial coordinate. This gives a circle through the origin. Consequently, the second term, which is proportional to $\sinh(r)$ in the denominator of the force, introduces distortions in the circle transforming it into an oval or an ellipse.

2.3 Newton's impact method

The proof of the expression for the centrifugal force in the hyperbolic plane,

$$CF = \frac{v^2}{\sinh(r)}$$ (2.30)

can be accomplished by Newton's impact method.[4] There is a serious error in quoted papers that the editors refused to correct claiming that the paper was 'too old' for such a correction to be of any interest to their readers. Errors are never too old to correct! The centrifugal force is given there as $v^2/\tanh(r)$ instead of (2.30).

The circular path is discretized by an n-sided regular polygon. The polygonal path is inscribed in a circle of radius r with center S, as shown in (2.5).

Keeping in mind that we are dealing with uniform circular motion, like Newton did. The only difference is that we are going to replace the distance in the source by the hyperbolic distance and not by the radius of curvature, as Newton did.

Two adjacent sides of the polygon are denoted by BC and CD. At point C where the collision occurs, a tangent EF is drawn in (2.5) (a). $EF \perp SC$ and $\triangle SCD$ is

---

[4]R L Lamphere, "Solving the non-euclidean uniform circular motion problem by Newton's impact method", *Math Mag* **83** (2010) 366.

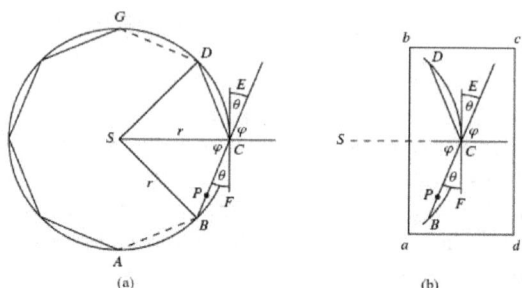

Figure 2.5: A particle at P will collide with a hyperbolic circle at point C. The image is taken from R L Lamphere, "Solving the non-euclidean uniform circular motion problem by Newton's impact method", *Math Mag* **83** (2010), p. 366.

congruent to $\triangle SCB$ it follows that $\angle BCF = \angle DCE$. This is the equality between the angle of incidence, and the angle of reflection, both equal to the angle θ. The rebound speed v is constant after each collision as the particle makes its way around the polygon.

Relative to the orthogonal axes CS and CE in (2.5) (b) the particle's moment pre-collision is $mv(\sin\theta, \cos\theta)$, while post-collision it will be $mv(-\sin\theta, \cos\theta)$. Consequently, the change in the particle's linear momentum will be $mv(-2\sin\theta, 0)$. This produces an impulse

$$f\Delta t = -2mv\sin\theta.$$

Referring to (2.5) (a) again, we apply the hyperbolic law of sines to $\triangle SCB$ to get

$$\frac{1}{\sinh(r)} = \frac{2\cos\varphi}{\sinh\Delta s}$$

where φ is the complement of θ, $\Delta s = BC$, and $\sin(\pi - 2\varphi) = 2\sin\varphi\cos\varphi$.

Introducing the law of hyperbolic sines into the impulse gives

$$f\Delta t = -\frac{\sinh(\Delta s)/\Delta s}{\sinh(r)} \cdot mv\Delta s,$$

where we have multiplied and divided by Δs. Summing over all n-sides of the polygon

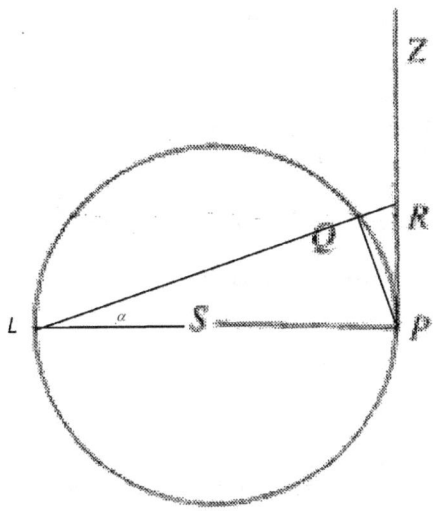

Figure 2.6: A Euclidean circle

gives a total impulse of

$$f \cdot n\Delta t = -\frac{\sinh(\Delta s)/\Delta s}{\sinh(r)} \cdot mvn\Delta s$$

In the limit $n \to \infty$, and both Δt and Δs tend to zero in such a way that $n\Delta t = T$, and $n\Delta s = L$, with $v = L/T$, we arrive at the hyperbolic centrifugal force (2.30) with unit scale factor, $R = 1$.

Newton's direct method was to determine the force given the trajectory. He did this in the Euclidean plane for a particle traveling in a circle and at constant speed v around a center S, as shown in (2.6).

In the Euclidean plane, the formula

$$RP^2 = RL \times RQ \tag{2.31}$$

holds where RP is the tangent to the circular motion, and RL is the hypotenuse of the right $\triangle PLR$. So the problem was to show that $CF = mv^2/SP$ was the correct expression for the centrifugal acceleration.

In contrast to Lamphere's interpretation [5] we incline the secant RL to the point L so that LP is the diameter of the circle. The secant makes an angle α with respect to the diameter. On the strength of Thales theorem the angle $\angle LQP$ is a right angle. Since $\angle RLP = \alpha$,

$$\angle LPQ = \angle QPR = \frac{\pi}{2} - \alpha,$$

the complementary angle. Newton then had recourse to Galileo's expression for the velocity and acceleration

$$RP = v\delta t \tag{2.32}$$

$$QR = \frac{1}{2}\frac{f}{m}(\delta t)^2, \tag{2.33}$$

where δt is an infinitesimal time increment. In the law of sines for $\triangle LRP$,

$$\frac{\sin\alpha}{\sin(RP)} = \frac{1}{\sin(LR)} = \frac{\cos\alpha}{\sin(LP)},$$

the last equality gives

$$\cos\alpha = \frac{\sin(LP)}{\sin(RP)}.. \tag{2.34}$$

Applying the law of sines, this time to $\triangle RQP$, gives

$$\frac{\sin\alpha}{\sin(QP)} = \csc(RP).$$

Eliminating $\sin\alpha$ between the two laws of sines results in

$$\sin(QR) \times \sin(LQ) = sin^2(RP), \tag{2.35}$$

which is the elliptic analogue of (2.31).

---

[5]R L Lamphere, "Solution of the direct problem of uniform circular motion in non-euclidean geometry" *Amer Math Momthly* **109** (2002) 650.

Now, introducing the kinematical relations (2.32) and (2.33) into (2.35), and going to the limit $\alpha \to 0$ result in

$$\sin\left(\frac{1}{2}\frac{f}{m}(\delta t)^2\right) = \frac{\sin^2(v\delta t)}{2\sin(r)},$$

where the radius $r = SP$, of the elliptic circle of circumference, $2\pi \sin(r)$, has been introduced. The circumferences of elliptic circles are decisively smaller than their Euclidean counterparts.

In the limit of infinitesimally small times the sines of the arguments containing δt become the arguments themselves so that the terms in δt cancel out leaving [6]

$$f = \frac{mv^2}{\sin(r)}.$$

The analogous expression for the centrifugal force in the hyperbolic plane of constant curvature is

$$f = \frac{mv^2}{\sinh(r)}$$

with a unit scale factor.

Similarly, it will not due to merely replace r by $r\tanh(r/R)$, and claim that

$$r = R\tanh^{-1}\left(\frac{\ell}{1+\varepsilon\cos\theta}\right),$$

where ℓ is the semi-latus rectum, and ε is the eccentricity, is an ellipse in the hyperbolic plane of constant curvature. [7]

In the hyperbolic plane, the distance between points (r, θ) and $(r, \theta + d\theta)$ is

$$ds = \sinh(r)d\theta,$$

in contrast to its Euclidean analog $ds = rd\theta$. Instead of the circumference given by $2\pi r$, it is given by $2\pi\sinh(r)$, as Gauss well knew. So for large values of r the circumference is much greater than in Euclidean analog.

---

[6]It was incorrect of Lamphere to conclude that r should become $\tan(r)$ in the elliptic plane of constant positive curvature.

[7]S. Stahl, "Planetary orbits in constant curvature planes".

For large u, the Binet equation (2.27) becomes

$$u'' + u = \left(2u - \frac{1}{2u}\right)^3 = 8u^3 + 3u^2 - 6u - \frac{1}{8u^3}. \tag{2.36}$$

The principal term is an inverse-fifth force, followed by an inverse-fourth, inverse-cube, and finally a linear restoring force. Thus, the inverse-fifth force and the Hookean force are at extremes. This is interesting because both forces belong to the class of dual forces. Whereas, the inverse-fifth force is its own dual, the dual of the Hookean force is Newton's inverse-square law. The distinction regards the position of the source, whether it is at the center or at the focus, respectively.

To understand the appearance of a cusp, we need look no further than the cardioid ,

$$r = a + b\cos\theta,$$

with a and b constants. The metric is

$$ds^2 = dr^2 + (r - a)^2 d\theta^2,$$

leading to a Newtonian slope of

$$\frac{1}{(r - a)}\frac{dr}{d\theta} = \pm\tan\theta.$$

The Binet equation is

$$u'' + u = 2(b^2 - a^2)u^3 + 3au^2. \tag{2.37}$$

For $a = 0$, the inverse-fifth law gives a circle passing through the origin. With $b > a$, the force is attractive, and a cusp forms. The cusp disappears when $a > b$, implying the force has become repulsive. The last term in (2.37) is reminiscent of the Einstein correction.

In fact, if we integrate (2.37) by u', and integrate we get

$$u'^2 = C - u^2 + 2au^3,$$

where C is an arbitrary constant of integration. With $a = \mu/c^2$, this describes the deflection of light, if we choose C appropriately and the inverse-square of the impact

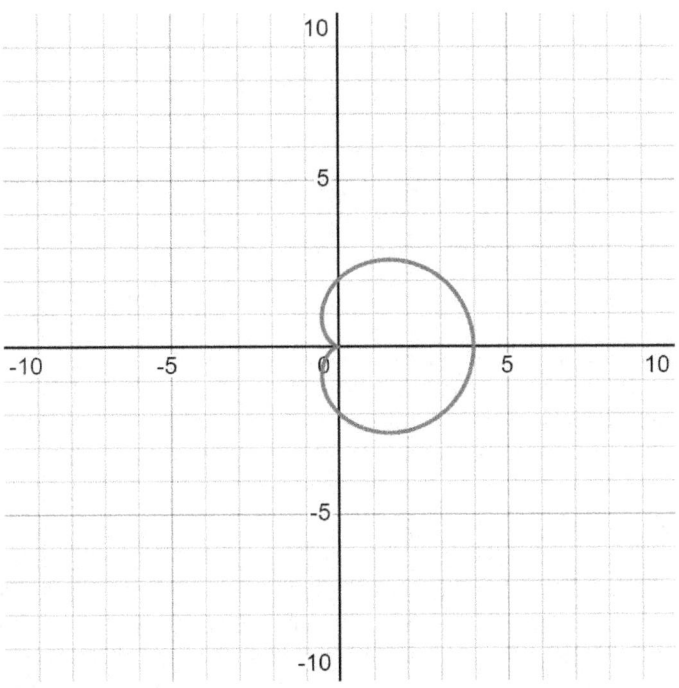

Figure 2.7: fig: Einstein's modification for the deflection of light without the inverse-square law still produces a closed orbit, but with a cusp, in contrast to the conclusions reached in all relativity books.

parameter. [8] This is another example where the 'correct' numerical value is left to a completely arbitrary constant of integration. The prime example is the exterior solution to the Schwarzschild problem.

However, there is a fly in the ointment in that the Binet equation (2.37) with $a - b$ *still* leads to a closed trajectory as the (2.37) shows.

With $C = \Delta^{-2}$, where Δ is the impact parameter, or the trajectory's closest approach to the source, the deflection of light calculated by Einstein's full theory of general

---

[8]C Moller, *The Theory of Relativity* (Oxford U. P., 1952), E. (45) p. 352.

relativity is $\Delta\theta = 4a/\Delta$ where $a = \mu/c^2$.

2.4 Newton's theorem on the algebraically non-integrable curves

Concerning Newton's theorem on the non-integrability of closed curves, a cusp in a curve indicates that the function is multi-valued, and this non-smooth behavior makes the curve non-integrable. In other words, the curve lacks a continuous first derivative at the cusp point, which prevents the existence of a smooth parametrization.

In this regard it is interesting to compare $y = \pm\sqrt{1 - x^2}$ with $r = \pm\cot^{-1}(A(\cos\theta + \sec\theta))$ as $A \to 0$. The former is a two-valued algebraic function representing two semi-circles. In the latter case, the two semi-circles for $A = 1$ develop cusps at the end points making it algebraically non-integrable because the first derivative becomes indeterminate at those points. In addition to circles, ellipses and hyperbolas are algebraic curves. However, they are non-integrable because they are transcendental, and cannot be solved by algebraic functions.

Kepler's equation, $M = E - \varepsilon\sin E$, between the mean anomaly, M (a parameterization of time) and the eccentric anomaly, E, is non-integrable. From the time of Newton, its solution has been sought in a power series in the eccentricity, ε, which is convergent for $|\varepsilon| \le 0.662743\ldots$.

It is not necessarily the appearance of singularities of cusps that make a curve algebraically non-integrable, but, rather, the nature of its transcendentalness. Arnold [9] who considers the oval $y = \pm\sqrt{x^2 - x^3}$ as being locally algebraically integrable shown in (2.8). Through a reparameterization which introduces the tangent of the angle of inclination, the area integral $\int y dx$ can be determined algebraically. If we limit the sweeping out of the ray through the nodal point to single swings, the area can be

---

[9]V I Arnold, *Huygens and Barrow, Newton and Hooke* (Birkhauser, 1990), p 86

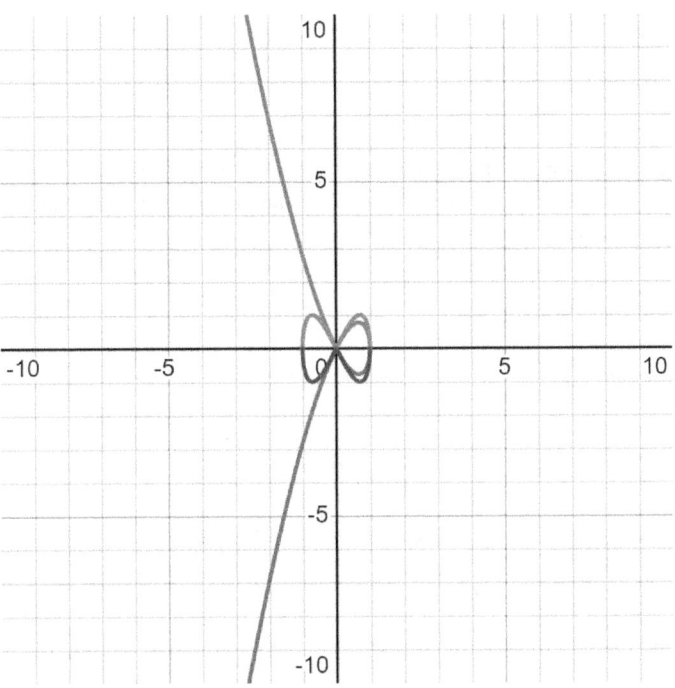

Figure 2.8: Oval and non-Bernoulli leminiscate having nodal points but are nevertheless algebracially integrable

calculated each time so that it is locally algebraic.

Newton's theorem predicts that the total area of a bounded self-intersecting closed curve, like the non-Bernoulli lemniscate, is locally algebraically integrable if for no other reason that the area of the two lobes cancel one another when their signs are taken into account. By deforming one of the lobes, the areas don't cancel and they are no longer locally algebraically integrable. The two closed curves are show in (2.8).

2.5 Generalized Ampère's laws and the pseudosphere

Thus far, we have obtained the Joukowski ellipse, corresponding to a one-soliton solution [10] closed trajectories in the non-Euclidean plane by modifying the G metric
coefficient in the metric

$$ds^2 = E dr^2 + G d\theta^2,$$

keeping the E coefficient unity. With E space dependent, we may write the metric as

$$ds = \sqrt{\left[\frac{E}{G} \left(\frac{dr}{d\theta} \right)^2 + 1 \right]} \sqrt{G}\, d\theta = \sqrt{\left(z^2 + 1 \right)} \sqrt{G} d\theta = \frac{\sqrt{G}}{\cos\theta} d\theta,$$

since

$$z = \sqrt{\left(\frac{E}{G} \right)} \frac{dr}{d\theta} = \pm \tan\theta,$$

is Newton's slope.' The term $\sqrt{(1 + z^2)}$ is basically the correction for the curve not
having standard arc length, $\sqrt{G}\, d\theta$.

It will prove convenient to write the metric in terms of the potential, $\Phi(r)$, as

$$ds^2 = \left(1 + \Phi'^{\,2} \right) dr^2 + r^2 d\theta^2,$$

transferring all the non-Euclidean dependence to the E coefficient of the metric. The
circles $r = $ const. are the parallels, and $d\theta = $ const. are the meridians on the surface
of revolution.

The radius of curvature, ρ, in terms of Φ will be given by

$$\rho = \frac{\left(1 + \Phi'^2 \right)^{3/2}}{\Phi''} \cdot r \frac{\sqrt{\left(1 + \Phi'^2 \right)}}{\Phi'}$$

Set $\Phi' = p$, and if we have a curve whose radius of curvature is proportional to the
normal, $r(1 + p^2)$. And if the radius of curvature is n times the normal

$$n \frac{(1 + p^2)^2}{p p'} = \pm r(1 + p^2),$$

---

[10]C Rogers & W K Schief, *Backlund and Darboux Transformations* Cambridge U P, 2002).

where the ambiguity in sign that each of the two lengths may be measured in two
different directions. If the radius and normal run in same directions, we choose the
positive sign. The differential equation reduces to

$$\frac{pp'}{1+p^2} = \frac{n}{r}.$$

Integrating gives

$$1 + \Phi'^2 = \left(\frac{R}{r}\right)^{2n},$$

upon reverting to the original variable, where R is a constant of integration which we
will set equal to unity. The radius of curvature will therefore be a power law,

$$\rho = -\frac{1}{nr^{2(n-1)}}.$$

The Gaussian curvature,

$$\mathcal{K} = \frac{\Phi'\Phi''}{r(1+\Phi^2)^2} = -nr^{2(n-1)} = 1/\rho.$$

will be negative if $n > 0$, and $\mathcal{K} = -1$ if $n = 1$.

Now, the equations of motion is derived from the Lagrangian,

$$\mathcal{L} = \frac{1}{2}\left(\frac{\dot{r}^2}{r^{2n}} + r^2\dot{\theta}^2\right) \tag{2.38}$$

beginning proportional to the time rate of change of the square of the distance interval.
The Euler-Lagrange equations

$$\frac{d}{dt}\left(\frac{\partial \mathcal{L}}{\partial \dot{r}}\right) - \frac{\partial \mathcal{L}}{\partial r} = F$$

$$\frac{d}{dt}\left(\frac{\partial \mathcal{L}}{\partial \dot{\theta}}\right) = 0,$$

where F is some external force, and the second equation expressed the conservation
of angular momentum for the cyclic coordinate, θ. The second equation asserts the

conservation of angular momentum, $L = r^2\dot{\theta} = $ const., while the first equation gives the force as

$$r\ddot{r} - n\dot{r}^2 - r^{2(n-1)n}L^2 = Fr^{2n+1}.$$

for $n = 1$, and $n = 1/2$, we get

$$\frac{1}{r^3}\left\{r\ddot{r} - \dot{r}^2 - L^2\right\} = F$$

$$\frac{1}{r^2}\left\{r\ddot{r} - \frac{\dot{r}^2}{2} - \frac{L^2}{r}\right\} = F, \tag{2.39}$$

respectively. There is no problem in recognizing the second equation as the Weber force with an angular term tacked on. Whereas in both equations, the centrifugal force is L^2/r^3, the force term in the first equation would be inverse-cube, while in the second, inverse-square. The former is related to Newton's revolving orbits, while the latter is the inverse-square law. This would mean that the total Weber force, and the resultant force in the first equation would be zero.

The soliton solutions arise when it is observed that the second derivative of

$$r = \ln\tan\left(\frac{\theta}{2}\right),$$

with respect to r, considering θ, a function of r, satisfies the sine-Gordon equation

$$\theta_{rr} \equiv \frac{d^2\theta}{dr^2} = \sin(2\theta) \tag{2.40}$$

This means that we should associate particle solutions to the Joukowski ellipse. But the Joukowski ellipse do not imply that we are on a pseudospherical surface of revolution. For that to be the case we will find it necessary to modify the E coefficient in the metric, keeping G flat. In the case $n = 1$,

$$\Phi(r) = \pm\int\frac{\sqrt{1-r^2}}{r}dr$$

$$= \pm(\sqrt{1-r^2} - \text{sech}^{-1}(r)), \tag{2.41}$$

which is the curve of a tractrix in two quadrants. The other two quadrants are obtained by $r \to -r$, and thus the surface of revolution is the pseudosphere shown in (2.9).

38

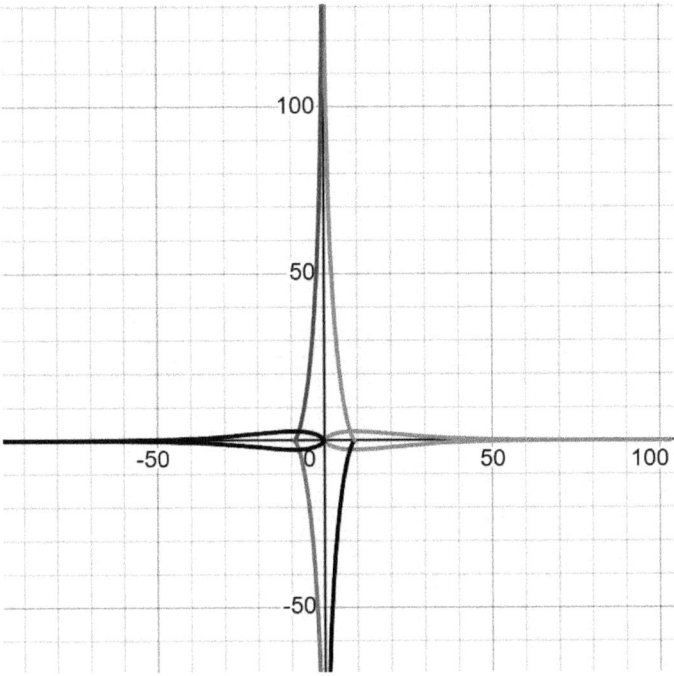

Figure 2.9: The pseudosphere obtained from Φ, and the Joukowski ellipses obtain by setting
$r = \sin \psi$.

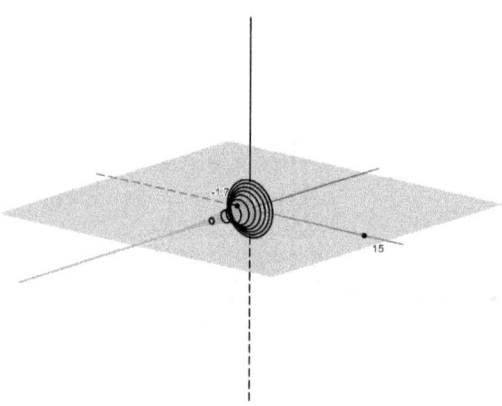

Figure 2.10: The bugle surface is one of constant negative curvature.

The surface of revolution is shown in 2.10. It does not fill all of Euclidean three-space.

Defining ψ by the circle $r = \sin \psi$, it is observed that

$$\frac{1}{r}\frac{dr}{d\psi} = \pm \cot \psi,$$

as the angle between the tangent to the curve and the curve itself in Newton's expression for his slope. It is just the equation of a circle, as one of the two choices in Newton's definition of his slope.

Inserting this into the expression for the potential, (2.41) results in

$$\Phi = \pm \left(\cos \psi - \ln(\csc \psi - \cot \psi) \right)$$
$$= \pm \left(\cos \psi - \ln \left| \tan \left(\frac{\psi}{2} \right) \right| \right)$$
$$= \pm \left(\cos \theta - \tanh^{-1}(\cos \theta) \right), \tag{2.42}$$

which are the two symmetrical Joukowski ellipses in (2.9). This highlights the intimate relation between the pseudosphere, in a hyperbolic plane of constant curvature, and the Joukowski ellipses.

Solitons can indeed be interpreted as moving on surfaces of constant negative curvature, such as the pseudosphere. This is because the negative and constant curvature of these surfaces allow for the stable propagation of solitons without distortion. Since there is a $1 : 1$ correspondence between the tractrix and Joukowski ellipse (2.42) their points of discontinuty must be related. The corresponding surface of revolution of the tractrix is a bugle surface. This surface cannot be extended beyond its rim to former a larger surface in Euclidean space because one of its principal curvatures tends to infinity while the other zero at the rim.

That is, if we consider the analytic description $u, r(u)$, where r is the solution to the differential equation $r' = r/\sqrt{c^2 - r^2}$, where the tangent line reaches the horizontal axis exactly at a distance c from the origin, the curve, $u = \sqrt{c^2 - r^2} - \operatorname{csech}^{-1}(r/c)$, will tend to zero as $r(u) \to c$, the principal curvatures are $-r'/c$ and $1/cr'$ so that the curvature $\mathcal{K} = -1/c^2$, as deduced previously. Thus the first derivative, $r' \to \infty$ as

$u \to 0$. Thus, the rim of the bugle displays the same discontinuous behavior as the cusp at the trailing edge of the Joukowski ellipse . Since boundaries in the hyperbolic plane correspond to points at infinity, where parallel railroad tracks will be seen to converge, these points will be inaccessible to anyone living on the hyperbolic plane because it will take him/her an infinite amount of time to reach the boundaries. This would tend to imply the the cusp at the trailing edge of the Joukowski ellipse is a vestige of the non-Euclidean plane that subsists in the complex conformal plane in Euclidean geometry.

The sine-Gordon equation has the form of the equation of motion of a simple pendulum—albeit with the wrong sign. And like the simple pendulum, it can be cast in terms of a Weierstrass function, which describes a point particle in a cubic potential. Since there can be three real roots, no single root can be globally stable and particle can make transition between these roots. The roots may be ordered according to the value of the constant energy, E, in the energy conservation equation,

$$\theta_r^2 + V(\theta) = 2E,$$

where the potential is $V(\theta) = \cos(2\theta)$, which has the characteristic form in Fig(2.11)). There are no solutions for negative energies less than -1; oscillatory solutions between -1 and $+1$; and continuous rotating solutions for energies greater than 1.

By a transform of the form $\cos\theta = a + b\wp$, where a, related to the total energy E, and b are two constant to chosen appropriately, the conservation of energy equation can be brought into the form

$$\wp'^2(z) = 4\wp^3(z) - g_2\wp(x) - g_3,$$

where $\wp$ is the doubly periodic Weierstrass elliptic function which is a function of the possibly complex variable, z, and g_2 and g_3 are constants. When they are real, and all three roots are also real, the half-period corresponding to the largest root is real, while that of the smallest root is imaginary. When two of the roots coincide, the Weierstrass function degenerates into a simply periodic function , expressible in terms of trigonometric or hyperbolic function [11] It is not our intention to pursue the relationship between the particle nature of surfaces of revolution of constant negative curvature, we turn our attention to the latter.

---

[11] See, for example, G Pastras, *The Weierstrass Elliptic Function and Applications in Classical and Quantum Mechanics* (pringer, 2020).

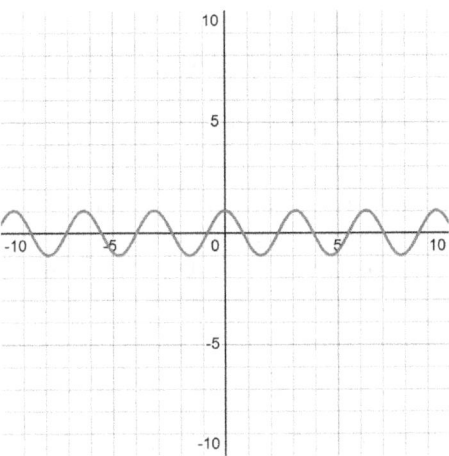

Figure 2.11: The simple pendulum potential

Unlike a normal ellipse, the Joukowski ellipse has a cusp, indicate of an abrupt change
in motion. If we associate the rotation of a particle in a normal ellipse. the soliton like
behaviour of the Joukowski ellipse would be indicate of multi-particles that interact at
the cusp that lead to a decisive change in the motion.

Epicycles show cusps; even a square is a possible resultant orbit of some finite combination of epicycles. indexepicycles However in order that the number of epicycles
remain finite, there must be cusps at the vertices of the square. In the limit that the
number of epicycles tends to infinity, the cusps will disappear.

In order to determine the coefficients in the second equation of (2.39). Weber argued
that particles in parallel current elements do not travel along the same straight line,
but diverge or converge exponentially according to

$$\ddot{r} = \frac{\dot{r}^2}{r}.$$

According to the first equation, this can be achieved by when the force is inverse-cube
since it would make the last term on the lhs and the force the same power of r. When
this condition is introduced in the second equation, the second derivative is eliminated

making the two terms comparable. The term with the negative sign represented the force between parallel current elements, and that with the plus sign the force between longitudinal elements. Ampère's experiments showed that the absolute value of their ratio was $1/2$, and this left one undetermined parameter in the equation. [12]

However, reducing the second-order equation to a first-order one to fix the coefficients it is inaccurate to return to a second-order equation and make believe the condition imposed by Weber never existed! For the value of the coefficients in the second-equation would only be valid under the condition imposed by Weber in which the inverse cube-force is balanced by the centrifugal force, $L^2/r^3 = F \propto 1/r^3$.

2.6 Gravitational Forces on the surface of a pseudosphere

Apart from being a model of the hyperbolic plane introduced by Eugenio Beltrami in the mid nineteenth century, the double cone, obtained by sticking together two pseudospheres at their ends, found interest as an uphill roller defying gravity more than a century and a half earlier.

William Leybourn in his recreational volume, *Pleasure and Profit,* published in 1694

describes how a double cone when placed on two inclined rails will actually roll uphill when the angle of inclination of the ramp upon which the double cone is placed, and free to roll, is large.

For it then behaves as a double cone defying gravity, while as it decreases the double cone becomes essentially a cylinder and rolls downhill. So even though the center of gravity moves downward, the double cone will ascend the ramp at large angles.

---

[12]L Hecht, "The significance of the 1845 Gauss-Weber correspondence" *21st Century* (1996) 22-43.

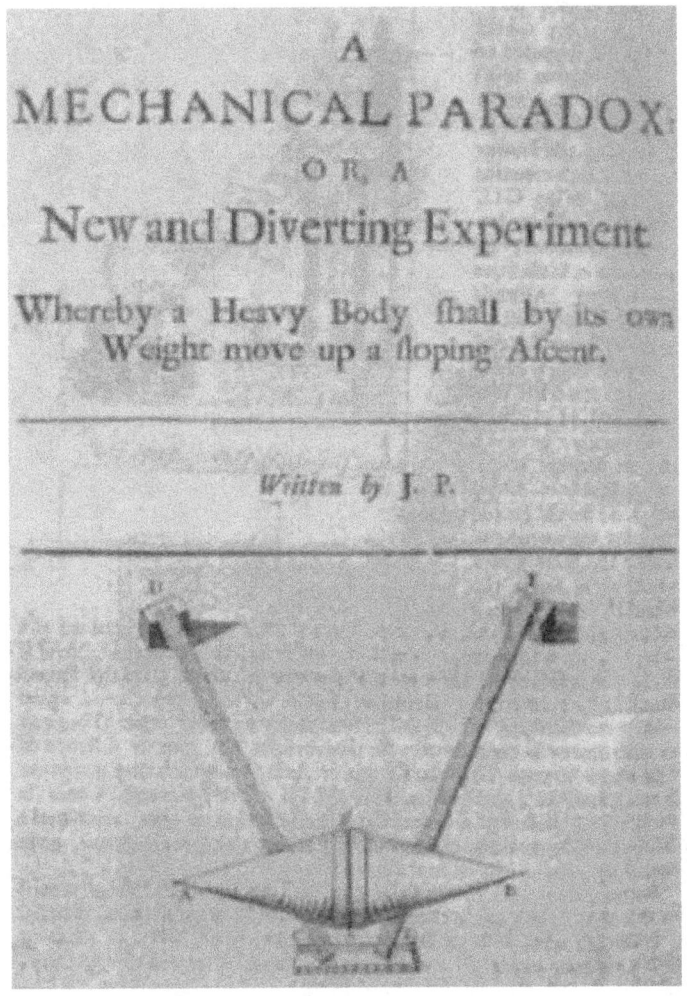

Figure 2.12: The cover of Leybourn's volume.

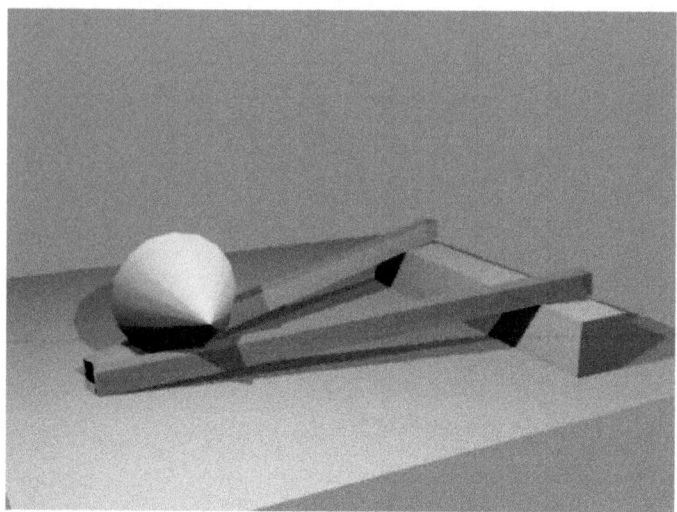

Figure 2.13: Double cone as an uphill roller defying gravity.

This could have been used as an argument on how geometry and gravity interact, although in this case it would give a surprising result of "gravitational repulsion". So it would be interesting to see what are the gravitational interactions that a body will display when constrained to move on the surface of a pseudosphere.

General Relativity goes to great lengths to show how masses can create indentions in a stretched membrane, such as a trampoline. Mathematically speaking, we can consider the embedding from Euclidean space, onto a curved space. However, since the pseudosphere has constant negative curvature, the hyperbolic plane is too big for $3D$ Euclidean space. If we consider the half-plane model of hyperbolic geometry in $2D$ Euclidean space, we are forced to consider a sector of the hyperbolic plane of with $2\pi c$, where c is some constant in the x-direction and less than c in the y-direction. But rather than being a hindrance, it is an attribute since the cut-off, c, will provide a criterion of validity of hyperbolic geometry as we shall now see.

Consider the metric of a flat cylinder of radius, r, and height, h:

$$ds^2 = dr^2 + r^2 d\varphi^2 + dh^2.$$

Furthermore, let us introduce the constraint.

$$dr^2 + dh^2 = d\rho^2,$$

where ρ is logarithmic distance

$$\rho = R \ln \left(\frac{r}{R} \right)$$

where R will play the role of the constant cut-off, c, above.

Introducing the constraint into the metric results in

$$dh^2 = \left(1 - \frac{r^2}{R^2} \right) dr^2,$$

which we have previously called Φ in our expression for the E coefficient of the metric. So that taking the negative square root gives the equation of a tractrix,

$$\frac{dh}{dr} = -\frac{\sqrt{1 - \frac{r^2}{R^2}}}{r}.$$

and evolving the tractrix along its long axis gives a pseudosphere.

Consequently, the constraint on the cylindrical metric gives rise to a $2D$ metric of constant curvature,

$$ds^2 = d\rho^2 + R^2 \exp\left(2\frac{\rho}{R}\right) d\varphi^2.$$

We recall that

$$ds^2 = d\rho^2 + R^2 \sinh^2\left(\frac{\rho}{R}\right) d\varphi^2$$

describes the hyperbolic plane of constant negative curvature..

Now, Newton's inverse square law asserts the force in $3D$ Euclidean space as proportional to the inverse of the surface area $4\pi r^2$ of a sphere. The natural extension to the hyperbolic plane would be to replace the surface area element by $4\pi R^2 \sinh^2(\rho/R)$, as the metric of the hyperbolic plane of constant negative curvature would suggest. But, this would give the area swept $\mathcal{A}$ out by a double radius, 2ρ, in an arc $R\Delta\varphi$ as

$$\Delta\mathcal{A} = \Delta\varphi R^2 \int_0^{2\rho} e^{x/R} dx = \Delta\varphi R^2 \exp\left(2\frac{\rho}{R} - 1\right)$$

or

$$\frac{d\mathcal{A}}{d\varphi} = R^2 \sinh(\rho/R)e^{\rho/R}.$$

The rate of change of the area with respect to time would be

$$\frac{d\mathcal{A}}{dt} = \frac{d\mathcal{A}}{d\varphi}\dot\varphi = R^2 \sinh(\rho/R)e^{\rho/R}\dot\varphi = 2L,$$

where L is the conserved angular momentum.

Thus, instead of Newton's law for the acceleration,

$$a = \frac{\mu}{R^2 \sinh^2(\rho/R)}$$

in the hyperbolic plane, which would reduce to his law in R^3 in the limit $R >> \rho$, we

now have

$$a = \mu \frac{e^{-\rho/R}}{R^2 \sinh(\rho/R)}.$$

In the asymptotic limit, $R >> \rho$, we get

$$a \to \mu \frac{e^{-\rho/R}}{R\rho},$$

which is an inverse-law rather than being an inverse-square law!

This is what MOND (modified Newtonian dynamics) predicts for rotational curves of spiral galaxies of extremely low centripetal accelerations. In that theory everything is lumped into the single parameter,

$$a = \frac{\sqrt{\mu a_0}}{\rho}.$$

A comparison with the above asymptotic value of the acceleration gives Milgrom's parameter as

$$a_0 = \frac{\mu}{R^2} e^{-2\rho/R}. \tag{2.43}$$

The exponential factor results from a shielding by other masses. Since a_0 is proportional to the total luminosity, L, of the galaxy, (2.43) would predict an exponential decay of the luminosity with logarithmic distance, ρ, or an inverse-square law for the distance, r,

$$a_0 = \mu/r^2.$$

Thus, it is the logarithmic scale of distance which is used on the pseudosphere causing an exponential decrease in luminosity, whereas it decays as a function on the inverse-square of the distance in terms of r, as one would normally expect.

As a final remark, we note that Weber's force has the associated potential

$$\Phi = \sqrt{r(1-r)} + \tan^{-1}\left(\sqrt{\frac{r}{1-r}}\right)$$

has the shape of hemisphere as shown in (2.14)

The double egg transform, $r = \sin^2 \psi$, gives a Newtonian slope

$$\frac{1}{r}\frac{dr}{d\psi} = \pm 2 \cot \psi.$$

However, unlike the case $n = 2$, this case, $n = 1$, is not associated with a surface of negative constant curvature; rather, the curvature is negative but non-constant $\mathcal{K} = -1/2r$.

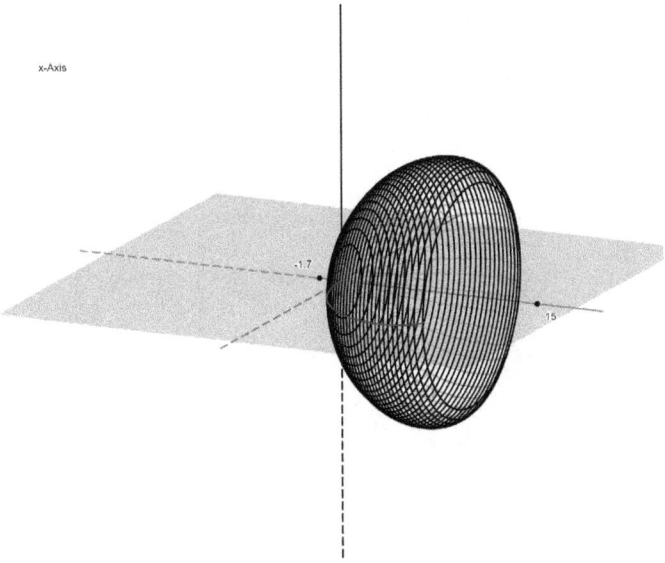

Figure 2.14: Surface of revolution for the Weber force is a hemispherical shape although it is non-constant and non-positive.

2.7 Optical Newtonian Gravity

We first rewrite the equation of angular momentum (2.2) as:

$$\sin \alpha = \frac{\Delta}{nr} \tag{2.44}$$

where we replace the *tangential* velocity v by the index of refraction, n, according to:

$$v^2 = \mu n^2, \tag{2.45}$$

where $\sqrt{\mu}$ is the constant of proportionality. This is because the tangential velocity behaves essentially as the inverse of the velocity of translation. [13] Consequently, we have $\Delta^2 = L^2/\mu$.

The radius of curvature ϱ in (2.1) can be expressed as:

$$\varrho \sin^3 \alpha = \left(1 + z^2 - \frac{r}{2} \frac{d}{dr}(1 + z^2) \right)^{-1}. \tag{2.46}$$

Introducing (2.44) results in

$$\varrho = -n^2 r / \Delta \cdot dn/dr. \tag{2.47}$$

The product the radius of curvature and the cube of the sine of the aberration angle α is:

$$\varrho \cdot \sin^3 \alpha = -\frac{\Delta^2}{r^2 \cdot dn^2/dr}$$

The condition that the left-hand side be constant implies that the index of refraction be of the form:

$$n^2 = \frac{2}{r} - \frac{1}{a}. \tag{2.48}$$

In optics, a lens with an index of refraction of the form (2.48 is known as an Eaton lens whose rays are shown in Fig (2.15).

---

[13] See §3.3.

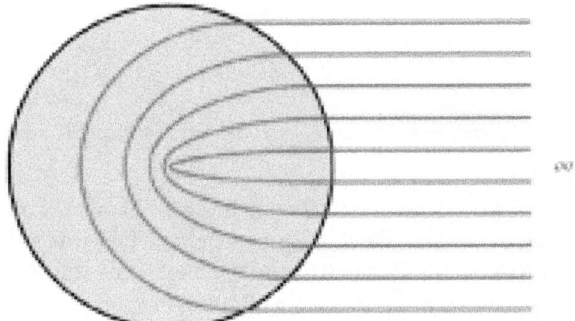

Figure 2.15: The Eaton lens where rays coming from infinitely far away enter a medium of refractive index (2.48), valid for $r \leq 2a$, or twice the semi-axis of the ellipse. They are then bent and return in the same direction they entered making for a perfect cat's eye.Image taken from U Leonhardt & T Philbin, *The Geometry of Light* (Dover, New York, 2010).

Consequently,

$$\varrho \cdot \sin^3 \alpha = \Delta^2 = \frac{L^2}{\mu} = a(1 - \varepsilon^2).$$

Introducing this back into (2.4) gives precisely Newton's inverse square law.

The equation of the trajectory,

$$u'' + u = \frac{\mu}{L^2}$$

is therefore an ellipse,

$$r = \frac{L^2}{\mu[1 + \varepsilon \cos(\vartheta - \vartheta_0)]}$$

where L^2/μ is the product of the semi-major axis a and $(1 - \varepsilon^2)$, with $\varepsilon < 1$ as its eccentricity.

The ellipse lies in the plane, and taking our cue from optics, if we were to raise our eyes to the sphere from which it is projected, we would find an index of refraction:

$$n = \frac{1}{a^2 + r^2},$$

(2.49)

which is Maxwell's fish-eye, [14] where a is the radius of the circular orbit.

The young, perspicacious, Maxwell wrote, in 1854, about the existence of such a medium as "possessing remarkable optical properties, [as] was suggested by the contemplation of the crystalline lens in fish and the method of searching for these properties was deduced from Newton's *Principia* Book I, Proposition VII". So someone else was reading Newton in the context of an application to optics!

The fish-eye mirror makes for a perfect lens, although it is rather peculiar in that it contains both the object and image in the same optical medium.

Luneburg [15] added a new twist to Maxwell's fish-eye when he found that the fish-eye performs a *stereographic projection* of the surface of a sphere onto a plane. It creates an illusion that light propagates on a sphere with a radius of curvature, $\varrho = 1/2\Delta$. Since a sphere has a constant positive curvature, $1/\varrho$, it represents a non-Euclidean medium where light travels along curves rather than rays. The stereographic projection produces Keplerian ellipses in the plane, as shown in Fig (2.16).

Light rays on the sphere travel on curved trajectories and those that follow the shortest path are those of great circle routes. Light rays emitted from one point meet at the antipodal point, thereby creating a perfect image. Maxwell's fish-eye is a little strange insofar as the object and image are confined to the same optical medium. Although it does not magnify it can reach an optical resolution that is independent of wavelength.

The reason why Maxwell referenced Proposition VII of the *Principia* is that it refers to a field of attraction that is the inverse fifth power of distance in a circle of radius a where light (gravity) is emitted from a point and focused at its point of inversion, as shown in Fig 2.17.

---

[14]J C Maxwell, *Cambridge & Dublin Math J* **8** (1854) 8.
[15]R K Luneburg, *Mathematical Theory of Optics* (Brown U, Providence R I, 1944) p 213.

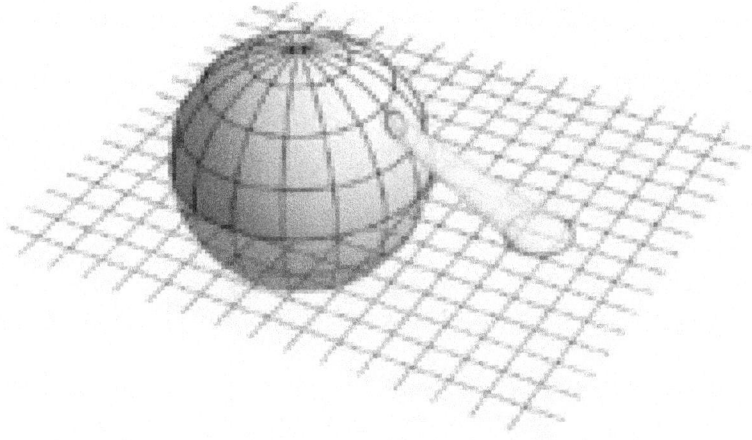

Figure 2.16: When light propagates a distance d in real space (the plane), it propagates a distance dr' on the sphere at the same time. The index of refraction, (2.49), is simply the ratio of these two distances. Ellipses on the sphere appear as ellipses in the plane. Taken from U Leonhardt & T Philbin *op. cit.*

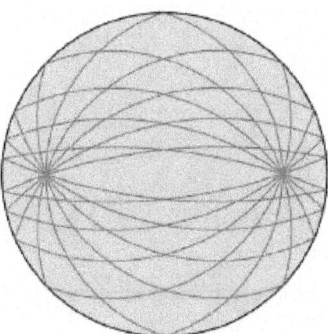

Figure 2.17: Perfect imaging can be achieved with Maxwell's fish eye where rays emitted from any point of the medium are focused at the antipodal point. Taken from U Leonhardt & T Philbin, *op. cit.*

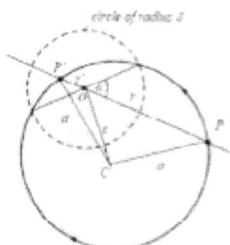

Figure 2.18: Maxwell's fish-eye where the center of the orbit, O, is displaced from the center of force which is located at the center of the circle whole circumference contains points P and P'.

According to Newton, the rays in Maxwell's fish-eye were the orbits of the planets, and he considered what would happen when the center of attraction did not coincide with the center of force. We should be impressed with the facility there is in going from heavenly bodies to rays of light.

The force is no longer central so angular momentum is not conserved. When the product

$$\varrho \cdot \sin^3 \alpha = \frac{\Delta^2 (a^2 + r^2)^3}{2r^3}$$

is introduced into (2.4) the force,

$$F = \frac{2\mu r}{(a^2 + r^2)^3},\qquad(2.50)$$

is asymptotically an inverse fifth power law.

Consider the case of a circular orbit of radius a as shown in Fig (2.18)

The equation of the path of light can be obtained from Fermat's principle of least time. It translates into requiring that the integral:

$$\int_{P_0}^{P_1} n(r)\sqrt{\left(1 + r^2 \vartheta'^2\right)}\, dr\qquad(2.51)$$

between any two fixed points P_0 and P_1, where the prime denotes differentiation with respect to r, be an extremum. Since ϑ is a *cyclic* (aka ignorable, kinosthenic, etc.) variable, in that it does not appear in the integrand, we know from the start that there exists a first integral:

$$L = \frac{nr^2\vartheta'}{\sqrt{(1 + r^2\vartheta'^2)}}.$$

Solving for ϑ' results in:

$$\vartheta - \vartheta_0 = \int \frac{L\,dr}{r\sqrt{n^2 r^2 - L^2}},$$

where the angle ϑ_0 is an arbitrary constant of integration if the limits on the integral are left indefinite.

Now, the *tour de force* in solving the integral is the recognition that

$$\frac{L(1 + \rho^2)}{\rho\sqrt{\rho^2 - L^2(1 + \rho^2)}} = \frac{d}{d\rho}\sin^{-1}\left(\frac{L}{\sqrt{1 - 4L^2}}\frac{\rho^2 - 1}{\rho}\right),$$

where $\rho = r/a$. Thus, it must be true that

$$\frac{r^2 - a^2}{r\sin(\vartheta - \alpha)} = \frac{r_0^2 - a^2}{r_0\sin(\vartheta_0 - \alpha)}$$

for whatever value α may be. If $\vartheta_1 = \pi + \vartheta_0$ then

$$r_1 = a^2/r_0, \tag{2.52}$$

showing that it is an *inversion*. This says that all rays emanating from an arbitrary point P will converge to point P' on the line joining the two points that pass through O. The points of inversion lie on opposite sides of O such that the product of their distances from O is a^2. We have here what is called in optics, an 'absolute' instrument in which the imaging is an inversion.

2.8 Conversion of Forces

Armed only with the properties of an ellipse and Euclid's *Elements*, Newton succeeded in establishing the inverse square law for the center of force located at a focus of an

Figure 2.19: A one pound bank note where the artist put Sun at the center of the ellipse instead of at a focus as Newton's diagram in the background clearly indicates. The Sun should be moved to a focus, or Hooke should appear instead of Newton. This is why we added Hooke's image to the note.

ellipse, the inverse cubic law for a hyperbolic spiral, where the aberration angle, α, is constant, Maxwell's fish-eye, or a displaced center of force or Archimedes spiral, for which the radius of curvature is constant, and, last but not least, Hooke's law of force.

There is a noticeable absence of an inverse fourth law, which will turn out to be Einstein's contribution to the correction of a Keplerian ellipse, and the inverse seventh law which corresponds to a Cassini oval, or in the degenerate case to a Bernoulli lemniscate.

We will find that the inverse-fourth law is already included in the inverse seventh law by the existence of *dual* force laws. It answers the question as to what integer inverse distance law, when expressible in polar coordinates, are expressible in terms of elliptic functions. It turns out, rather mysteriously, that there are only two pairs, $[-5, -5]$, which is its own dual, and $[-4, 7]$, whose closed curves are obtained by slicing a torus rather than a conic, as in the case of the third pair $[1, -2]$.

It is rather ironic that the Bank of England should put Newton on their one pound note with the Sun at the center of the elliptical orbits of the planets, which was Hooke's idea, his archrival. Fig (2.1) can clearly be seen in the background in Fig (2.19).

Newton went to great lengths to distinguish the nature of the two forces that would result.

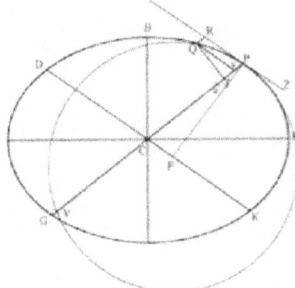

Figure 2.20: Newton's revised diagram for Proposition X in the 1713 revised edition of *Principia*. Taken from Brackenridge, *op. cit.*

In the second edition of *Principia*, which appeared in 1713, Newton revised his diagram for his Proposition X to the one in Fig (2.20).

Newton establishes the fact that whereas the force directed from the focus F of the ellipse follows the inverse square law, the force directed from the center C is proportional to the distance CP to the point P of the ellipse. The line PV, (alias SY not shown in the diagram) is constructed so that it is perpendicular to both the tangent PR to the ellipse and the conjugate diameter DK.

A chord PV of the circle of curvature at P is drawn, and Newton uses the known properties of an ellipse to establish $CP \cdot PV = 2CD^2$. The sine of the complementary angle between the distance from the focus to the point P and the center to the same point is proportional to their ratio PF/CP.

The radius of curvature is

$$\varrho = CD^2/CP \cdot \sin \alpha,$$

so that

$$\varrho \sin^3 \alpha = \frac{CD^2}{CP} \frac{FP^2}{CP^2}.$$

59

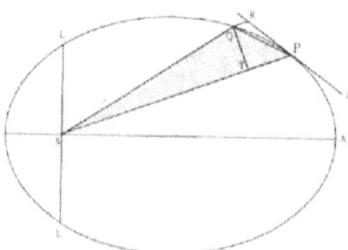

Figure 2.21: As the point Q tends to the point P, the ratio QR/QT^2 tends to the inverse of the line L which is the *latus rectum* of the ellipse.

When this is multiplied by CP^2, and the inverse taken, the law of force,

$$F_c \propto \frac{CP}{(CD \cdot PF)^2} \propto CP, \tag{2.53}$$

is proportional to CP since $FP \cdot CD$ is the circumscribed area that, according to Lemma 12, is constant for any ellipse.

Consequently, (2.53) is Hooke's linear law for any force operating from the center of the ellipse. Hooke's law is to be paired with Newton's inverse square law which can be given the following simplified derivation.

A tangent to the motion of P, RPZ is drawn in Fig (2.21).

Since QR is the deviation of the planet from its elliptical trajectory, Newton uses the second law to set the force proportional to the ratio of the distance QR and the square of the time interval. From Kepler's law, the time is proportional to the area swept out, in this case the triangle SPQ. That area is proportional to be product of the base SP and the altitude QT. Thus, the force is

$$F \propto \frac{QR}{(QT \cdot SP)^2}.$$

Newton then demonstrates in Problem 3 and Proposition XI that as $Q \to P$ the ratio QR/QT^2 becomes proportional to the inverse of the latus rectum L, which is a constant for any ellipse. Consequently, $\propto 1/SP^2$, which is Newton's inverse square law.

Strange things happen when we shrink or magnify geometrical objects. Consider the equation of an ellipse:

$$r = \frac{\varepsilon p}{1 + \varepsilon \cos \vartheta},$$

where ε is the eccentricity and p the distance from the minor axis to the latus rectum. The product εp is the semi-latus rectum, which in the Kepler problem is L^2/μ, where L is the angular momentum, not to be confused with the L in Newton's diagram. Thanks to Newton, we know that the force field is inverse square. But, what happens when we say take its square root?

Consider:

$$r = \frac{r_0}{\sqrt{(1 + \varepsilon \cos \theta)}}.$$

Taking the derivative in time gives

$$\dot r = \frac{\varepsilon \sin \vartheta r^3}{r_0^2} \cdot \frac{L}{r^2},$$

where we used Kepler's second law, $\dot\vartheta = L/r^2$. Squaring results in

$$\dot r^2 = -\frac{(1 - \varepsilon^2)L^2 r^2}{r_0^4} - \frac{L^2}{r^2} + \frac{2L^2}{r_0^2}.$$

With

$$ds^2 = dr^2 + r^2 d\vartheta^2,$$

as the invariant arc length, the centrifugal force is one-half the derivative of

$$v^2 = \dot r^2 + \frac{L^2}{r^2}$$

with respect to r. We thus find the expression for the centrifugal force as

$$F = -\frac{(1 - \varepsilon^2)L^2}{r_0^4} r, \tag{2.54}$$

which is Hooke's linear law. For an ellipse the direction of the force is in the *opposite* direction to r, while for an hyperbola, it is in the same direction of increasing r. So,

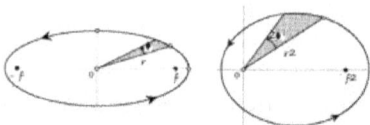

Figure 2.22: Squaring (2.55), known as the Bohlin map, transforms a Hookean ellipse with its attracting point located at the center, with foci $\pm f$, into one that has one of its foci at the origin and the other the square of the original foci, f^2. Distances between the attracting center to a point on the orbit, r, has been squared, and angles, ϕ swept out have been doubled. The cross term in the square of the Joukowski transform shifts the left focus to the origin.

merely by taking the square root of the polar equation of an ellipse we get the transmutation of Newton's inverse square law into Hooke's linear law for a force exerted on a spring that distinguishes between the direction of the force and the direction of deformation. The transformation between a Hookean and Newtonian ellipse upon square is shown in Fig (2.22).

The dual laws are best understood by using complex variables. The square of a Hookean ellipse,

$$w = z + \frac{1}{z},$$
(2.55)

becomes a Newtonian ellipse with the center at one of the foci,

$$w^2 = z^2 + \frac{1}{z^2} + 2.$$
(2.56)

Equation (2.55) is known as a Joukowski ellipse, and when it is squared, (2.56) it is again a Joukowski ellipse, except that it is shifted by 2.

In other words, any ellipse with a focus at the origin is the square of an ellipse whose center is at the origin. [16] The displacement of the center by the transform $z \mapsto z^2$ transforms the orbit of a linear field into an orbit of the inverse square field.

Even more can be said when we consider the *inverse* of the Joukowski transform (2.55)

---

[16]T Needham, "Newton and the transformation of force," *Am Math Monthly* **100** (1993) 119-137.

when we write it as $w = (1/2)(z + 1/z)$. For its inverse will satisfy:

$$\tilde{w} \equiv w^{-1} = \frac{2z}{1 + z^2},\qquad(2.57)$$

since z is its own inverse. Equation (2.57) is the isomorphism that takes two hyperbolic models, the Klein and Poincaré disc models of the hyperbolic plane into one another. Geodesics consist of chords in the former, and circular arcs that cut the rim of the disc orthogonally. Whereas the former is not conformal, the latter is.

The isomorphism (2.57) can be written equivalently as:

$$\frac{1 + \tilde{w}}{1 - \tilde{w}} = \left(\frac{1 + z}{1 - z}\right)^2.\qquad(2.58)$$

Taking logarithms of both sides gives the relation between *hyperbolic* length segments in the Klein and Poincaré models. Equation (2.58) is analogous to the transformation, $w = z^2$, that relates the laws of force of Hooke and Newton. This points to some deep-seated connection between hyperbolic geometry, which we shall return to in the sequel, but now we want to address the topic of *dual* laws.

2.8.1 Dual Laws

Suppose that the complex variable, z, obeys Hooke's law:

$$\frac{d^2z}{dt^2} \pm k^2 z = 0,\qquad(2.59)$$

Then its square $w = z^2$ will satisfy Newton's law:

$$\frac{d^2w}{d\tau^2} = -c\frac{w}{|w|^3},$$

where k and c are constants, and t and τ are two different time scales.

From the law of areas stating that equal areas will be swept out in equal times,

$$\frac{|z|^2 d\vartheta}{dt} = \text{const.}, \qquad \frac{|w|^2 d\vartheta}{d\tau} = \text{const.},\qquad(2.60)$$

it follows that $d\tau/dt = |w|^2$ so that $d/d\tau = z^{-1}|z|^{-1}d/dt$. Consequently,

$$\frac{d^2w}{d\tau^2} = -2\frac{|\dot{z}|^2 + k^2|z|^2}{z|z|^3} = -2Ez^{-1}|z|^{-3},$$

where

$$E = (1/2)(|\dot{z}|^2 + k^2|z|^2)$$

is the total energy for the positive sign in (2.59), and hence $c = 4E$. This is usually referred to as Bohlin's theorem who discovered it in 1911, although records show that Kasner formulated it two years earlier.

The negative sign in (2.59) corresponds to the repulsive case of a hyperbola, as we saw in (2.54) where $\varepsilon > 1$. Now, the energy is

$$E = (1/2)(|\dot{z}|^2 - k^2|z|^2),$$

and the dual field is attractive, or repulsive, depending on whether the total energy, E, is positive or negative.

The question naturally arises as to whether there are other dual pairs. Using the equation of the orbit, a general expression for the force can be written as: [17]

$$L^2 u^2 \left(u'' + u \right) = F\left(\frac{1}{u}\right). \tag{2.61}$$

One would expect that the force in (2.61) can be expressed in a power series in u. The constant term, μ/L^2, the inverse of the semi-latus rectum, gives the equation of the orbit for an ellipse, and, hence this coincides with Newton's inverse square law.

The linear term corresponds to Cotes' spirals , whose potential energy is $V = -\mu/r^2$. The equation of the trajectory is

$$u'' + \left(1 - \frac{\mu}{L^2}\right) u = 0. \tag{2.62}$$

If $L^2 > \mu$, the motion will never terminate; the radial coordinate has a single minimum value, an apsis, and the orbit branches off on either side. On the contrary when $L^2 < \mu$, the particle will spiral into the origin, as shown in Fig (2.23).

The relation between Cotes' spiral and black holes come to mind. But, black holes exist, as convention dictates, when the outer Schwarzschild metric is extended beyond its domain where the radius falls below its event horizon radius, known as the Schwarzschild radius.

[17] J.M.A. Danby, *Fundamentals of Celestial Mechanics* (Macmillan, New York, 1962) Eq (4.3.2).

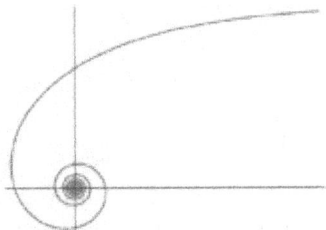

Figure 2.23: A black hole with a 'naked' singularity? Cotes' spiral of 'fatal attraction when the angular momentum is overpowered by gravitational attraction. Sketch taken from U Leonhardt & T Philbin, *op. cit.*

2.8.2 Einstein's Modification of the Keplerian ellipse

The Schwarschild velocity was first derived by Droste [18], and confirmed by Hilbert. [19] Its square is:

$$\dot{r}^2 = \left(1 - \frac{2\mu}{c^2 r}\right)\left[1 + A\left(1 - \frac{2\mu}{c^2 r}\right)\right], \tag{2.63}$$

where A is an arbitrary constant. In order that the velocity vanish as $r \to \infty$ requires that $A = -1$. The invariant line element gives the square of the velocity as:

$$v^2 = \dot{r}^2 + \frac{L^2}{r^2} = \frac{2\alpha}{r}\left(1 - \frac{\alpha}{r}\right) + \frac{L^2}{r^2}, \tag{2.64}$$

where $\alpha = 2\mu/c^2$.

The centrifugal force is $v \cdot v'$, where the prime stands for differentiation with respect to r; it is given by:

$$F_S = -\frac{\mu}{r^2}\left(1 - \frac{\alpha}{r}\right)\left(1 - 3\frac{\alpha}{r}\right) - \frac{L^2}{r^3}. \tag{2.65}$$

For purely radial motion, it has been claimed that gravitational *repulsion* occurs in the range

$$1 > \alpha > \frac{1}{3},$$

---

[18]J. Droste, Het zwaartekrachtsveld van een of meer lichamen volgens de theorie van Einstein (E J Brill, Leiden, 1916).
[19]D. Hilbert, Die Grundlagen der Physik, *Math Ann* **92** (1924) 1.

for it appears from (2.65) that the force changes sign there.

Yet, the general equation for the trajectory is:

$$u'' + \left(1 + 4\frac{\alpha}{L^2}\right) u = \frac{\mu}{L^2} + 3\frac{\alpha^2\mu}{L^2}u^2. \tag{2.66}$$

The term that would lead to gravitational repulsion has been incorporated into the linear term on the left-hand side. It does not even 'threaten' a Cotes' in spiraling. The quadratic term is the Einstein correction to a Keplerian ellipse that seemingly gives the correction to the motion of the planet (in this case Mercury) that appears to be slowly rotating in space. This is due to the fact that the coordinate defining the line of apsides is continually changing.

Does the "3" in the modified Einstein equation tell us something? If we attach the centrifugal and Coriolis forces to the force of gravity, the acceleration reads

$$a = \frac{\mu}{r^2} + \Omega \left(\omega \wedge \omega \wedge r - 2v \wedge \omega\right)$$

where $\Omega = \mu/c^2 r$ is the analog of the magnetic permeability, and v is the velocity in the rotating frame. Replacing v by $\dot{r}$, and projecting the acceleration in the radial direction add on a factor of 2 to the centrifugal acceleration,

$$a \cdot \hat{r} = \frac{\mu}{r} + 3\frac{\mu L^2}{c^2 r^4},$$

where we introduced the angular momentum , L. Now, dividing by $h^2 u^2$, we add this to the right hand side of the Binet equation to get precisely Einstein's modification of a Keplerian orbit.

According to the Wikipedia article on the Lense-Thirring effect, that, unlike the Coriolis force, the last term in the acceleration is "not fictional, but due to "frame-dragging" introduced by a rotating body. However, "frame-dragging" is not something new that was discovered by GR, but, rather, can be explained by traditional, fictitious forces.

Einstein's modification of the Keplerian ellipse (2.78), and also Newton's inverse square law, (2.6), implicate an inverse fourth power law. It is enlightening to consider Newton's inverse problem, where we want to determine the law of force given the orbit.

In this case, we consider the general equation of a limaçon (which in French means snail):

$$r = a + b \cos \vartheta, \tag{2.67}$$

where a and b are constants. Taking the time derivative of (2.67), introducing Kepler's areal law, squaring, and eliminating the angular dependence finally yield:

$$\dot{r}^2 = (b^2 - a^2)\frac{L^2}{r^4} + \frac{2aL^2}{r^3} - \frac{L^2}{r^2}.$$

The square of the total velocity is:

$$v^2 = \dot{r}^2 + \frac{L^2}{r^2} = (b^2 - a^2)\frac{L^2}{r^4} + 2\frac{aL^2}{r^3}.$$

The central force,

$$F = \frac{1}{2}\frac{dv^2}{dr} = 2(b^2 - a^2)\frac{L^2}{r^5} + 3\frac{aL^2}{r^4},$$

is a combination of inverse fourth and inverse fifth laws. For $a/b < 1$, we get a looped limaçon, shown in Fig (2.24), which is attractive.

For $b/a < 1$, the inverse fifth law becomes repulsive, and either a dimpled, $1 < a/b < 2$, or a convex, $a/b > 2$, limaçon results. The gap in the rate of angular precession of the line of apsides in Mercury's orbit implies that $b = a$, and $a = \mu/c^2 L^2$. In this case, the limaçon reduces to a cardioid, shown in Fig (2.25).

A cardioid results by tracing out a point on a circle as the circle winds around a fixed-circle of equal radius, $a = b$. In general limaçons are formed when a circle rotates around another circle, $a \neq b$, of unequal radii. This brings to mind an epicycle that moves on a larger circle known as a deferent. The law of force is an inverse fifth power law.

It was known since the time of Le Verrier that the perihelion of Mercury precesses at a slightly faster rate than can be explained by Newtonian gravity. During the last decade of the nineteenth century it was fairly popular to propose modification of Newton's inverse square law to account for the slow rotation. Typical proposals were for an advance of the orbital perihelion per revolution of the order of

$$\kappa \pi \left(\frac{\mu}{L c_G}\right)^2, \tag{2.68}$$

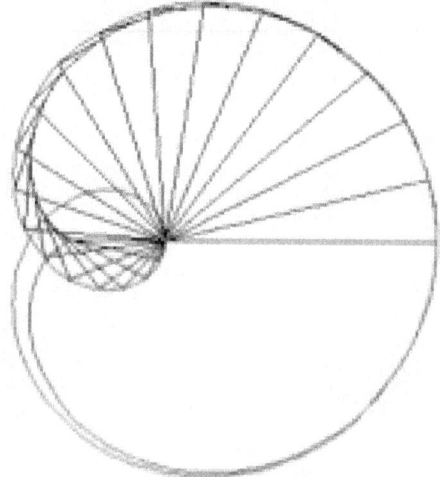

Figure 2.24: A looped limaçon having an attractive inverse fifth law since $a/b < 1$. It is also known as a Cayley sextic being a pedal curve of a cardioid. Its involute is a nephoid which is a special case of an epicycloid.

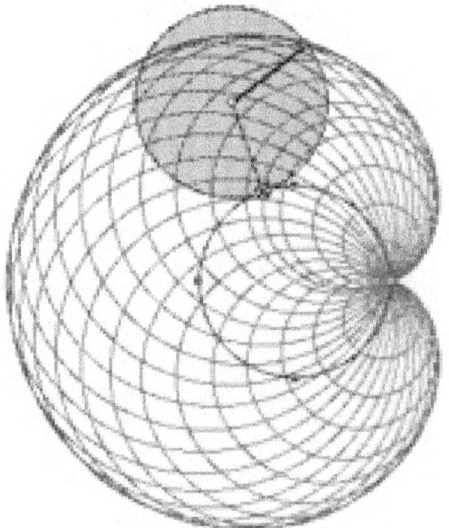

Figure 2.25: A cardioid can be constructed by rolling a fixed circle around a larger circle. Image taken from J Wilson, UGA "Cycloids."

where κ is a fitting parameter, and c_G was the speed at which gravity propagates. Using Kepler's III, $\mu = a^3 \omega^2$, where a is the semi-major axis of the ellipse, $P = 2\pi/\omega$ is the known period, the rate of angular change of the line of apsides could be determined.

In order to give the missing $43''$ arcsec per century, it was necessary to set $\kappa = 3$, and to take c_G as the speed of light, lending support that gravity travels at such a speed.

Furthermore, if the expression for the Schwarzschild force, (2.65), is correct, and that of the resulting equation for the orbit, the ratio that appears as a square in (2.68) should again be squared, leading to the fourth power of the ratio!

The reason is that there is an alternative method that begins with the metric,

$$\frac{\dot{r}^2}{1 - \alpha/r} + \frac{L^2}{r^2} + 1 = c^2 \dot{t}^2 (1 - \alpha/r), \tag{2.69}$$

with the invariants

$$r^2 \dot{\vartheta} = L \qquad\qquad (1 - \alpha/r)\dot{t} = \text{const.}$$

The force is

$$F_S = -\frac{\mu}{r^2} - 3\frac{\mu L^2}{c^2 r^4} + \frac{L^2}{r^3}. \tag{2.70}$$

With $F_S = \ddot{r} = -L^2 u^2 u''$, the equation of the orbit becomes

$$u'' + u = \frac{\mu}{L^2} + 3\frac{\mu}{c^2} u^2. \tag{2.71}$$

In comparison with (2.66), this gives the rate of change of the line of apsides, (2.68), for the planet of Mercury. We should also point out that (2.70) can never become positive so that any reference to repulsive gravity is a consequence of the method of solution, rather than being one of substance.

Even if the expression were correct, it could be considered as gap fitting at best, because it does not offer a physical explanation for the rate of angular change of the line of apsides. We shall now show that the fourth inverse power law corresponds to the orbit of a limaçon so that the orbit of Mercury is a mixture of a predominantly elliptical orbit with a 'dash' of a limaçon that causes it to slowly rotate. It belongs the class of magical dual laws, and its dual is that of the inverse seventh power law whose orbit describes a Bernoulli lemniscate.

A conformal mapping, $w = z^\alpha$ can be looked upon as a relation between two different media with differing indices of refraction. A trajectory with energy, E, and potential, $V(z) = |dw/dz|^2$, possesses a dual trajectory with energy, $\tilde{V}(w) = -|dz/dw|^2$, and potential, $\tilde{E}$, where $E\tilde{E} = -1$.

If $n^2(z) = E - V(z)$ is the index of refraction of the z-medium, we have the optical analogy that

$$n^2(z) = E \left| \frac{dw}{dz} \right|^2 \tilde{n}^2(w),$$

where the index of the w-medium is

$$\tilde{n}^2(w) = \tilde{E} - \tilde{V}(w).$$

Since the areas $|w|^2$ and $|z|^2$ are different, we have every reason to suspect that the times $d\tau$ and dt it will take to sweep these areas out will be different. However, if the law of areas is not to be violated, we must have

$$|w|^2 \frac{d\vartheta}{d\tau} = \text{const.} \qquad |z|^2 \frac{d\vartheta}{dt} = \text{const.}$$

We can therefore choose to set

$$\frac{d\tau}{dt} = \frac{|w|^2}{|z|^2} \tag{2.72}$$

and use

$$\frac{d}{d\tau} = z^{-1}\bar{z}^{-1}\frac{d}{dt}$$

where the bar indicates complex conjugate to show that Newton's law,

$$\frac{d^2w}{d\tau^2} = -4E\frac{w}{|w|^3},$$

is equivalent to Hooke's law,

$$\frac{d^2z}{dt^2} + z = 0,$$

if $E = |\dot{z}|^2 + |z|^2$ is the total energy.

A generalization of Bohlin's theorem [20] has been given by Arnol'd [21] to the effect that central forces of degree a in the z-plane transform into those of degree $\tilde{a}$ in the w-plane under the transformation $w = z^\alpha$, where

$$(a + 3)(\tilde{a} + 3) = 4, \qquad \alpha = \frac{a + 3}{2}. \qquad (2.73)$$

Although these relation remain a physical mystery, they do point to the fact that if the exponent of the force law is an even integer, its dual must be an odd integer. There are only three such pairs:

$$[1, -2] \qquad [-4, -7] \qquad [-5, -5].$$

The inverse cube of Cotes' spiral is absent, and the inverse fifth law is its own dual. We conjecture that if one of these power laws appear in the expression for the force, the dual law will be absent. Moreover, they do not all refer to conic sections: the last pair includes Cassinian ovals which are obtained by slicing a torus, just like conics result from slicing a cone.

2.9 From Force to Metric

An indefinite metric in the plane of the form,

$$c^2 d\tau^2 = A(r)c^2 dt^2 - B(r)dr^2 - C(r)r^2 d\vartheta^2,$$

has two time scales. τ and t refer to the different areal times under the scale change of the areas.

---

[20] K Bohlin, Note su le problème des deux corps et sur une integration nouvelle dans le problème des trois corps, *Bull. Astr.* **28** (1911) 113-119.

[21] V I Arnold, *Huygens and Barrow, Newton and Hooke* (Birkhauser Verlag, Basel, 1990).

Dividing both sides by $d\tau^2$, the right-hand side can be considered as a lagrangian, $\mathcal{L}$. The variational equations:

$$\frac{d}{d\tau}\frac{\partial\mathcal{L}}{\partial\dot{x}} - \frac{\partial\mathcal{L}}{\partial x} = 0,$$

yield two first integrals of the motion:

$$A\dot{t} = \text{const}, \qquad (2.74)$$

and

$$C(r)r^2\dot{\vartheta} = L. \qquad (2.75)$$

Equation (2.74) appears as a generalization of the equivalence of two areal laws with different areas that occur in two different time increments, (2.72). It is usually referred to as the conservation of energy, and (2.75) is a generalization of Kepler's areal law.

Setting the arbitrary constant, $A\dot{t} = 1$ in (2.74) and $AB = 1$, we obtain the square of the total velocity as:

$$v^2 = \left(c^2 + \frac{L^2}{r^2}\right)\frac{\alpha}{r}. \qquad (2.76)$$

In comparison with the Newtonian expression for the energy, (2.76) contains an additional term which is responsible for the modification of the Keplerian ellipse. The form of the potential, commonly referred to at the Einstein potential, V_E, is shown in Fig (2.26).

Recalling that the force is $F = (1/2)dv^2/dr$, we get:

$$F = -\frac{\mu}{r^2} - 3\frac{\mu L^2}{c^2 r^4}.$$

And inserting this into the equation for the orbit of the trajectory, we find:

$$u'' + u = \frac{1}{L^2 u^2}(-F) = \frac{\mu}{L^2} + 3\frac{\mu}{c^2}u^2. \qquad (2.77)$$

In the case of light which travels on light cones, $d\tau = 0$, the equation of the orbit (2.77) reduces to:

$$u'' + u = 3\frac{\mu}{c^2}u^2. \qquad (2.78)$$

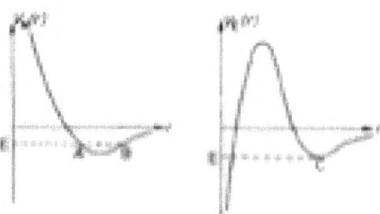

Figure 2.26: Comparison of the Newtonian potential, V_N, and Einstein's modification, V_E. Elliptic orbits exist between radii A and B in V_N. Similar orbits for V_E, where the extrema correspond to unstable and stable, C, circular orbits. Taken fom R. Sexl & H. Sexl, *Weisse Zwerge-scharze Locher* (Rowohit Taschenbuch, Hamburg, 1975) p 76.

This corresponds to the square of the total velocity as:

$$v^2 = c^2 + \frac{\alpha L^2}{r^3}.$$

The force is then

$$F = -3\frac{\mu L^2}{r^4}. \tag{2.79}$$

We remind the reader that eighteenth century natural philosophers divided the problem relating the force to the orbits as: the 'direct and inverse problems'. In the former, we are given the path and the center from which the force acts, and are asked to determine the force that is necessary to determine the motion. Whereas, in the latter, we are given the force and the force center, and are required to find the path. Equation (2.79) represents the inverse problem.

We have emphasized this distinction for the following reason. In GR, the force (2.79) is implicated in the deflection of light about a massive object, like the Sun. The term, $3\mu/c^2$ is considered as a very small perturbation, but with respect to what? If it pertains to a light ray, we should expect that the trajectory of a light beam is essentially a straight line. The effect of the extremely small perturbation $\epsilon = 3\mu/c^2$ is to alter the trajectory so that it bends the ray slightly, but in which direction?

We know from the above discussion that the force (2.79) corresponds to a closed path trajectory know as a cardioid (cf., Fig (2.25)). How then can ϵ be considered a small perturbation to a straight line $r = r_0 / \cos \vartheta$ that produces a small deflection as light approaches the Sun along an asymptotically straight line?

That the putative deflection, $\Delta = 4\mu/c^2 r_0$, gives a numerically acceptable result of $1.75''$ should be considered as extremely fortuitous, rather than as a confirmation of a theory. [22]

Realizing something amiss, Adler *et. al.* (ABS) [23] try to discredit the solution $u' = $ const., or what they consider $u = $ const. According to ABS, the solution "would describe light rays circling the attracting center at a fixed distance, $r_0 = 3\alpha/c^2$." In this case, ABS claim that
$$u'' + u = \epsilon u^2$$
only for $r_0 = \epsilon$, and such a constant solution "cannot be changed continuously into the more general equation (2.77) except at $r_0 = \epsilon$."

However, ABS fail to explain how a constant solution can be changed into a non-constant solution *at* $r_0 = \epsilon$. ABS write these solutions off as being "unstable and of no physical interest." But, how can
$$r = \frac{\mu}{c^2}(1 + \cos \vartheta)$$
be written off as unstable and of no physical interest?

ABS then go on to claim that "the general equation for light rays should be of second order so that rays through every point and in every direction are possible. This condition is fulfilled by [(2.77)], but not by $u' = 0$."

In general the contrary is true: [24] As long as there is a conservative field, (2.77) is not as useful as:
$$\dot{r}^2 = \left(2\frac{\mu}{c^2}r - 1 \right)\frac{L^2}{r^2}. \tag{2.80}$$

---

[22] See the *Introduction.*
[23] R Adler, M Bazin & M Schiffer, *Introduction to General Relativity* (McGraw-Hill, New York, 1965).
[24] Danby, *op. cit.* p 61.

In fact, (2.77) can, rather tediously, be recovered by differentiating (2.80). Its only saving grace is that the force can be expressed in powers of u.

In conclusion, we may ask why do we need perturbation theory when there is an exact solution no matter how small the parameter ϵ is?

And there is still a stronger argument here against the manipulations of GR. GR does not deal with forces at all, and even if it did what is there that would guarantee that the principle of superposition would hold? In fact, that principle is **not** valid in GR.

2.10 Curvature Determines Force

We have seen how majestically Newton employed the concept of curvature to determine central forces. To him, curvature was a manifestation of force, and not part of the woodwork. Almost three centuries later, Einstein got rid of forces entirely by associating curvature with geometry, only to bring them back when considering the work that they do.

Einstein's viewpoint was succinctly stated by Wheeler: "Spacetime tells matter how to move, matter tells spacetime how to curve." Yet, even without matter, spacetime is curved! When matter is placed in spacetime, it will sink to greater depths because of gravity, as seen in Fig (2.27).

So spacetime does not curve without mass, and its curving is due to the force of gravity driving it down. Without the force of gravity what else would cause the mass to sink?

This conundrum forced out gravitational energy from the energy-stress tensor which accounts for all other forms of energy other than gravitation. Rather, Einstein found that gravity was described by 'pseudo'-tensor t_ν^μ, which is not a tensor at all since it is composed only of the components of the metric tensor, and their first derivatives. Levi-Civita [25] was quick to point out that such 'invariants' under general coordinate

---

[25]T Levi-Civita, "On the analytic expression that must be given to the gravitational tensor in Einstein's theory," *Rendiconti Acc Lincei* **26** (1917); transl arXiv:physics/9906004.

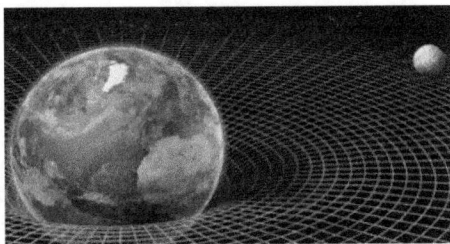

Figure 2.27: A mass like the Earth creates a dent in spacetime thereby attracting other masses. This is according to GR, but what causes the dent if it is not the force of gravity tending to pull the mass downward?

transforms do not exist! As Gauss well knew, you need second derivatives to tell you how things curve.

Consider the Laplacian, Δ, the divergence of the gradient field. It tells you how much the field changes from that of one of a steady variation that there would be in a divergence free flow.

2.11 Sectional Curvatures

With the advent of Riemann geometry, a rotationally symmetric metric with a conformal factor G will have all its sectional curvatures lying between

$$K_{12} = \frac{R^1_{212}}{g_{22}} = -\frac{G''}{G}, \tag{2.81}$$

and

$$K_{23} = \frac{R^2_{323}}{g_{33}} = \frac{1 - G'^2}{G^2}. \tag{2.82}$$

where the prime stands for differentiation with respect to the independent coordinate which is usually the radial coordinate, but it can be also time. Equations (2.81) and (2.82) run the gamut from *radial* to *tangential* curvatures.[26] . Sectional curvatures are

---

[26]cf. P Peterson, *Riemann Geometry* (Springer, New York 2006) p 72.

obtained by taking a two-dimensional slice through a given plane, and measuring the two-dimensional curvature of the slice.

For constant K_{12}, equation (2.81) has the form of Jacobi's equation. It describes how the geodesics, or shortest paths on the given surface, will spread or bunch up. If $K_{12} < 0$, the geodesics will spread out in the hyperbolic plane, while if $K_{12} > 0$, they will bunch up in the elliptic plane. For tangential curvature K_{23} will be a positive constant on S^2, while if negative and constant, it will describe the diffusion of geodesics on a hyperboloid.

If there are forces, there will be deviations from geodesic trajectories, and these local deviations can be expressed in terms of how volumes get distorted and twisted.

Newton showed that *tidal* forces are accounted for by μ/r^3, like that of Cotes'spirals. If we consider their action on a sphere, two components will have negative accelerations while the third has twice the strength, and acts to separate geodesics. The bulging of the earth at the equator, and its flattening at the poles are manifestations of tidal forces.

In a state of dynamic equilibrium, the sum of tidal forces vanishes. This immediately brings to mind Einstein's condition of emptiness, yet, here nothing is empty! According to Dirac [27], Einstein's condition of emptiness

$$R_{\mu\nu} = 0, \qquad\qquad (2.83)$$

constitutes a law of gravitation. 'Empty' here means that there is no matter present and no physical fields except the gravitational field. The gravitational field does not disturb the emptiness. Other fields do.

What is stated in words is not borne out by matter of fact.

Label the variables t, r, ϑ, and φ by the indices $0, 1, 2$, and 3, respectively. The Ricci

---

[27]P A M Dirac, *General Theory of Relativity* (Wiley, New York, 1975).

tensor , say in the 2-direction, is the sum of sectional curvatures:

$$\text{Ric}_2 = -K_{21} - K_{21} - K_{23} = \frac{F'G'}{FG} + \frac{G''}{G} - \frac{1 - G'^2}{G^2}, \tag{2.84}$$

where the factors F and G are metric coefficients in

$$d\tau^2 = F^2(r)dt^2 - dr^2 - G^2(r)d\sigma^2, \tag{2.85}$$

where

$$d\sigma^2 = d\vartheta^2 + \sin^2\vartheta d\varphi^2 \tag{2.86}$$

is the standard, constant curvature, metric on a sphere. We assume, furthermore, that
the metric is both static and spherically symmetric which are reflected in the fact that
the metric coefficients are only a function of the radial coordinate, r.

2.11.1 Is there a Gravitational effect analogous to electrodynamic induction?

The title comes from a little known article of Einstein (1912) in which he shows that a
mass of a system is modified by the presence of other matter by an amount GMm/Rc_0^2,
where m is the mass within a hollow sphere of mass M whose radius is R, and c_0 is the
speed of light distantly far away from the sphere. Here, the masses m and M are symmetrical, like two current elements in Ampère's law. But, instead of finding repulsion
between the two circuit elements, Einstein finds attraction between the masses.

This was Einstein's attempt to incorporate Mach's principle into his theory where inertia is not a property of either M or m, but rather exists because of the presence of
the other. Unfortunately, his later theory of GR does not satisfy Mach's principle for,
as we shall see in the next section, GR allows mass to exist in an otherwise empty
universe, and still manifests the property of gravitational attraction to what relativists
refer to as a 'test' mass.

Over forty years later, the English physicist Sciama raises the same question, apparently unaware of Einstein's 1912 paper. Sciama argues that the rest of the universe
determines the inertial frames, and, as such, inertia is not an intrinsic property of matter, but, rather, exists because of the presence of matter in the universe.

Again, this evokes an analogy with electrodynamic induction, whereby a circuit carrying current i' can affect another circuit of current i that is at a distance r from it. This

is the epitome of the dreaded 'action-at-a-distance,' which allows the propagation of disturbances to occur infinitely fast.

Relativity has failed to aid us in understanding the interaction of bodies that are well-separated in space. Ampère didn't need a constant speed of light in formulating his law that a force exerted by a complete circuit on a current element is at right angles to the element.

Only when static and motional forces are combined must we introduce limits on the propagation of disturbances. Recall that Weber found it necessary to introduce a constant which represented the number of Coulombs of charge that had to travel through a wire in a given amount of time in order that it be equal to one Ampèrian of current. In doing so, Weber found that the ratio of electrostatic to electromagnetic charges was roughly that of the speed of light.

How does sectional curvature play a role in determining the nature of the forces? First, the action must occur along the radial distance separating the two currents or two masses. Second, no non-local derivatives are considered, as Ampère did.

In the previous section we saw that all sectional curvatures lie between $-r''/r$ and $(1-r'^2)/r^2$, where the prime denotes the derivative with respect to the tubular element, s, or the azimuthal angle, φ. This is a result of Riemann geometry, which, in two-dimensions coincides with Gaussian curvature. Let us consider Gaussian curvature in greater detail.

Gaussian curvature can be expressed by his formula,

$$K = K_0 - \Delta \ln r, \qquad (2.87)$$

where K_0 can be thought of as the 'unperturbed' curvature, and Δ is the Laplacian.

Undoubtedly, the simplest way to alter a surface is by applying a conformal change. The Gaussian formula, (2.87), relating two Gaussian curvatures applies to surfaces that have densities, and has found applications in such diverse areas as Perelman's proof of the Poincaré conjecture. The radial coordinate r will in general represent a conformal factor and the second term in (2.87) represents the variance, or mean-square deviation in the expected value, ∇r. Obviously, this invites the comparison of the conformal factor with the partition function of statistical mechanics. [28]

The second limit on sectional curvature relates to that of a circle. Gaussian curvature K is related to the area of a circle of radius r by the lowest order correction,

$$\text{Area} = \pi r^2 - K\pi r^4/12 + \cdots$$

Since we are considering circular motion, the area is proportional to the square of the areal velocity:

$$\text{Area} = \pi \frac{(\text{Areal velocity})^2}{4c^2} = \frac{1}{4}\pi r^2 \left(\frac{v}{c}\right)^2,$$

where v is the tangential velocity to the orbit of radius r. The areal velocity is the rate at which the area is swept out by a particle as it moves along a curve, which is Kepler's II.

Combining the two expressions for the area gives a Gaussian curvature:

$$K/12 = \frac{1}{r^2}\left\{1 - \left(\frac{v}{2c}\right)^2\right\}, \tag{2.88}$$

Equation (2.88) is a modified inverse square law, and if the tangential velocity is constant, it can be integrated to determine a potential. That potential is what Schroedinger began with in his determination of the advance of the perihelion of Mercury. In his words, the potential "shall depend on the masses and the separation of two points in the same manner as does the Newtonian potential, [and] it shall be proportional to the square of the rate of change of their separation," v. [29]

Returning to our theme of the analogy between electromagnetic and gravitational induction, we recall Neumann's original definition of the vector potential as: [30]

$$\vec{A} = i' \int_{s'} \frac{\hat{r} \cdot d\vec{s}'}{r}\hat{r}. \tag{2.89}$$

---

[28] cf. B H Lavenda, *Statistical Physics: A Probabilistic Approach* (Dover, New York, 2016) Ch 4.

[29] Rather than going through a lengthy explanation, Schroedinger could have simply cited previous work by Reissner, who got it from someone else, who got it from Weber's potential for the interaction of two charges. See, *Mach's Principle* J Barbour & H Pfister eds (Birhauser, Boston, 1995)

[30] For a good introduction to Neumann's work, see P Graneau, *Ampère-Neumann Electrodynamics of Metals* 2nd ed (Hadronic Press, Palm Harbor FL, 1994).

This represents the *global* effect of a circuit s' carrying current i' at a distance r from another circuit element, where $\hat{r}$ is the unit normal pointing in that direction.

Coulomb and vector potentials had only a secondary role, if you could call it that, in Maxwell's field equations. The major players were the electric, $\vec{E}$ and magnetic fields, $\vec{B}$. But all that changed when Aharonov & Bohm came up with a physical effect in which the vector potential determines the phase difference of an electron circling a solenoid which is not single-valued due to a phase shift in the wave function By varying the field strength within the solenoid an interference pattern could be produced that is observed as fringe shifts on a detector – even though there is no magnetic field present.

This was heralded as a victory for relativity over action-at-a-distance theories, like that of Ampère's. For it implied that $\vec{A}$ had to be treated as a physical field, for otherwise, it would be necessary to consider that $\vec{B}$ creates the phase shift through action-at-a-distance.

Analogous to the definition of the vector potential, (2.89), the magnetic field is defined by:

$$\vec{B} = i' \int_{s'} \frac{\hat{r} \wedge d\vec{s}'}{r^2},\tag{2.90}$$

which replaces the dot product in the expression for $\vec{A}$ by the vector product and an extra power of r is found in the denominator because it is a field, and not a potential.

Since it is only a difference of replacing the dot product by the cross product, the distinction between potentials and fields vanish, and so does the difference between relativity and action-at-a-distance.

We now consider $i \ln r$ as a partition function, and the vector potential, (2.89), as the expectation value, obtained from the partition function through differentiation:

$$\vec{A} = \frac{i}{r} \frac{\partial \vec{r}}{\partial s},$$

where the infinitesimal tube length, s, plays the role of inverse temperature.

The variance, or mean-square deviation, is obtained through a second differentiation, this time with respect to s', and so we get:

$$\frac{d\vec{A}}{ds'} = i\hat{r}\left(\frac{1}{r}\frac{\partial^2 r}{\partial s \partial s'} - \frac{1}{r^2}\frac{\partial r}{\partial s}\frac{\partial r}{\partial s'}\right).$$

If we take the dot product of this expression and integrate round a close circuit, we obtain:

$$\oint \vec{r} \cdot d\vec{A} = -i'\oint \frac{1}{r}\frac{\partial r}{\partial s}\frac{\partial r}{\partial s'}ds' = i'\oint \frac{\cos \epsilon}{r}ds' \equiv -i'M,$$

which defines the mutual inductance, M, introduced by Neumann. Neumann defined the integrand as the "mutual electomagnetic energy necessary to separate two circuits to infinity"

The *non-locality* of Ampère's law is essential in the definition of mutual induction of the two circuits. It cannot be obtained by any relativistic argument. In fact, there is a problem to introduce time.

The vector potential depends non-locally on the two circuits that are separated by a distance, r. Its dependence on time is another story. The transition from Ampère to Weber required replacing the current elements ids and $i'ds'$ by relative velocity and the time increment, dt. This was carried out using Fechner's hypothesis for which currents of opposite charges travel at the same speed but in opposite directions. Without his hypothesis the squared terms v^2 and v'^2 would not cancel when

$$\frac{dr}{dt} = v\frac{\partial r}{\partial s} + v'\frac{\partial r}{\partial s'}$$

is squared. Only terms involving vv' should subsist in the final formula. [31]

So we have seen how curvature, in general, and Gaussian curvature, in particular, measure deviations from geodesic paths. The Laplacian measures the curvature or stress of the field, and tells you how much the field differs from its average value with respect to neighboring points. This is the content of Gauss's relation (2.87). This will change as we move about, and, so, it may be considered as a random variable. However, this random variable is not without its own probability distribution!

---

[31]Maxwell, *op. cit.* §849, p 428.

The Laplacian is the divergence of a gradient, and as such it tells you how much the rate of change of the field differs from the kind of steady variation that you would expect from a divergence-free flow. In a single dimension:

$$\Delta = \frac{\partial^2}{\partial x^2} = \text{curvature},$$

In $3D$, if the Laplacian of a function vanishes, that function is said to be harmonic and satisfies an averaging principle. A non-vanishing Laplacian measures deviations from such behavior, where the deviations are treated as average of random variables.

Again in $3D$, the area of a sphere goes as r^2 so that if its divergence vanishes the radial solution is proportional to $1/r^2$, which is either Newton or Coulomb. Whereas in $2D$, the area goes as r itself so that the field must be increasing in order for its divergence to vanish. The corresponding potential is $\ln r$, which is the electrostatic potential in $2D$. It is also the potential of dipole-dipole interactions in Ampère's theory of two interacting current elements.

Curvature also determines the quadrupole force that is produced at the current element $id\vec{s}$ in the direction of the current element, $i'ds'$. The force is:

$$\vec{F}_{\vec{s} \to \vec{s}'} = \frac{3ii'}{4\pi r^4} \left\{ (\hat{r} \wedge \vec{s}' - 2\hat{r}(\vec{s} \cdot \vec{s}') + 5\hat{r}(\hat{r} \wedge \vec{s}) \cdot (\hat{r} \wedge \vec{s}') \right\}, \qquad (2.91)$$

which is not entirely directed along the radial coordinate connecting the two current elements. Hence, it will not satisfy Newton's III.

The force expression can be simplified by using the vector relation:

$$(\hat{r} \wedge \vec{s}) \cdot (\hat{r} \wedge \vec{s}') = (\vec{s} \cdot \vec{s}') - (\hat{r} \cdot \vec{s})(\hat{r} \cdot \vec{s}'),$$

since from it we can derive the law of cosines:

$$\cos \epsilon = -\cos \vartheta \cos \vartheta' + \sin \vartheta \sin \vartheta' \cos \omega, \qquad (2.92)$$

in a spherical triangle.

The left-hand side of the cosine law is related to mutual induction, as we have just seen where ϵ is the angle between the two current elements. The longitudinal components of the force law are $\partial r/\partial s = \cos \vartheta$, and $\partial r/\partial s' = \cos \vartheta'$, whereas the parallel component is:

$$r\frac{\partial^2 r}{\partial s \partial s'} = \sin \vartheta \sin \vartheta' \cos \omega,$$

where ω is the angle between the two planes containing the two current elements.

The quadrupole force, (2.91), can thus be simplified to:

$$\vec{F}_{\vec{s} \to \vec{s}'} = \frac{3ii'}{4\pi r^4} \left\{ \left(2r \frac{\partial^2 r}{\partial s \partial s'} - 3\frac{\partial r}{\partial s'} \right) \hat{r} + \frac{\partial r}{\partial s} \vec{s}' \right\}. \tag{2.93}$$

The quadrupole force (2.93) tells us that the ratio of longitudinal to parallel components is 3 : 2, and points along the direction separating the two current elements. It also tells us that there is a component of the force in the direction of s'.

o In an attempt to generalize Ampère's law, Maxwell added on terms such as that found in (2.93), but, finally, concluded that Ampère's radial law was the most frugal in that it acts in the radial direction only. This is not true of higher order forces like the quadrupole force.

The total quadrupole force will contain a contribution $-\partial r/\partial s' \vec{s}$, which will then make up the Grassmann force. Its gravitational analog will be

$$\vec{F}_G \sim \frac{G\eta M}{c^2 r^3} \omega^2 r \left\{ (2\sin\vartheta \sin\vartheta' \cos\Omega - 3\cos\epsilon)\hat{r} + \cos\vartheta \vec{s}' - \cos\vartheta' \vec{s} \right\},$$

where ηM is the reduced mass, $M = m + m'$, the total mass, Ω is the angle between the two planes in which the masses are found, and the last two terms comprise the Grassmann force. For GW, the Coulomb gauge is used to get rid of the longitudinal components in the gravitational radiation field $\vec{g}$, which is the negative of the time rates of change of a gravitational vector potential, $\vec{A}$: [32]

$$\vec{g} = -\frac{\partial A}{\partial t} = \frac{G\eta M}{2c^2 r} \left(\frac{\omega r}{c} \right) \omega^2 r \left(\sin 2\theta \sin 2\omega t \hat{\theta} - 2\sin\theta \cos 2\omega t \hat{\phi} \right),$$

which is the tangential component of the force since it has undergone first-order aberration. The polarizations in directions, $\hat{\theta}$ and $\hat{\phi}$ are unequal, in contrast to Grassmann's

---

[32]R C Hilborn, "Gravitational waves from orbiting binaries without general relativity: a tutorial," arXiv:1710.04635v1, Eq (48).

force in which they are equal in magnitude. Moreover, there is no reason to discard the radial component of the quadratic force in (2.93), which would be the analog of Ampère's dipole term. And since an EM model of GWs is being presented, it should be the Grassmann force that should appear. We will come back to our discussion of whether GWs are longitudinal or transverse in §6.2.

Moreover, projecting (2.93) in the radial direction,

$$\hat{r} \cdot \vec{F}_{\vec{s} \to \vec{s}'} = \frac{3ii'}{2\pi}(-K),$$

where

$$K = -\frac{\Delta \ln r}{r} = -\frac{1}{r^2}\left(\frac{1}{r}\frac{\partial^2 r}{\partial s \partial s'} - \frac{1}{r^2}\frac{\partial r}{\partial s}\frac{\partial r}{\partial s'}\right).$$

The terms within the parenthesis tell us that the Gaussian curvature is made up of translational curvature, and a component of tangential curvature, which would be the sectional curvature had we added the Coulomb potential. This shows that there is absolutely no use to considering the Ricci tensor in such circumstances, which is an average of sectional curvatures in any given direction. Finally, that the curvature is negative is indicative of a *repulsive* force that exists between the two current elements.

The numerator of K is related to the variance of a random variable, Z:

$$\text{Var}(Z) = \left\langle Z^2 \right\rangle^2 - \langle Z \rangle^2,$$

where its expectation is:

$$\langle Z \rangle = \frac{\partial \ln r}{\partial s}. \tag{2.94}$$

If the independent variable were ϑ, the angle between the centripetal force and the force directed to the center of force, the expected value (2.94) would be the average slope of the trajectory. In view of (2.93), a non-vanishing expectation is responsible for a component of the force in a non-radial direction.

Taking the derivative of (2.94) gives:

$$\frac{\partial^2 \ln r}{\partial s' \partial s} = \frac{1}{r}\frac{\partial^2 r}{\partial s' \partial s} - \frac{1}{r^2}\frac{\partial r}{\partial s}\frac{\partial r}{\partial s'} = \frac{\cos \epsilon}{r^2} = \left\langle Z^2 \right\rangle - \langle Z \rangle^2, \tag{2.95}$$

where ϵ is the angle opposite the side ω in (2.92).

The variance (2.95) is always positive semi-definite. It also relates the variance to an isoperimetric relation that is comparable to the lower bound given by the Cramer-Rao equality in statistics.

In this manner, the area of a region bounded by a curve of length $\langle r \rangle$ is related to the curvature of the surface. Replacing the length by a geodesic, or length minimizing curve, converts the expression for the variance into an isoperimetric inequality that is analogous to the Cramer-Rao inequality. The isoperimetric inequality is related to the uncertainty in the curvature, and this allows us to employ statistical concepts that would otherwise be a purely geometric problem. Bounds on the curvature would imply bounds on the forces from which they are created.

Equation (2.95) also shows that the variance is related to magnetic induction, which is the double integral of $\cos \epsilon / r$ around closed circuits of ds and ds'. Thermodynamically, the variance is related to extensive quantities, like the dispersion in energy which is measured by the heat capacity of the system. So magnetic induction may be related to stored energy. Just like its time rate of change gives the electromotive force, so that rate of decrease in dispersion in energy can be related to power.

2.11.2 The Schwarzschild Metric

A good example of such a metric is the Schwarzschild metric , where $G(r)$ will turn out to be the radial coordinate itself that will be confined to the open interval $(2\mu, \infty)$.

Consider first the spatial part of (2.85). The tidal forces will then be related to the sectional curvatures by:

$$K_{02} = K_{12} = -\mu/G^3 \qquad \text{and} \qquad K_{23} = 2\mu/G^3. \qquad (2.96)$$

This results in the vanishing of the Ricci tensor component, (2.84), which is a classical Newtonian result in a non-classical setting. The presence of tidal forces is a testimony to the non-emptiness of the universe.

When the value of K_{23} in (2.96) is equated to (2.82) there results:

$$G'^2(r) = 1 - 2\mu/G(r). \qquad (2.97)$$

If $K_{23} = \rho = $ const., then

$$G'^2 = 1 - \rho G^2. \tag{2.98}$$

The other two sectional curvatures in (2.96) reduce to the Jacobi equation:

$$G'' = \rho G, \tag{2.99}$$

and describes the spreading to geodesics that is caused by the action of 'tidal forces', which are essentially due to the curvature of the non-Euclidean metric, and, therefore, do not imply any destructive behavior.

What we have essentially made is a transformation of Schwarzschild's 'outer' solution, (2.97), into his 'inner' solution, bypassing any mention of black holes. This is exactly the manner in which Schwarzschild presented his two solutions. In other words, *there is no meaning to consider distances $G < 2\mu$ in his outer solution.* This is why his inner solution, which has constant curvature, and is related to the well-known Beltrami metric, has, literally speaking, been swept under the carpet and completely forgotten.

The Beltrami metric is obtained by multiplying the spatial component of the Schwarzschild metric, (2.100) below, by the conformal factor $1/G'^2$, and, therefore, is conformally equivalent to it.

Had we considered the Ricci tensor, which is an average of sectional curvatures, we would not have been aware of the action of tidal forces, and it would have appeared that the universe is entirely empty. Why then consider an average when the individual terms which make up the average are related to the forces?

For instance, the third condition is:

$$K_{02} = \frac{F'G'}{FG} = \mu/G^3.$$

From this condition, we find:

$$F = (1 - 2\mu/G)^{1/2} = G'.$$

This could also have been obtained by the vanishing of the Ricci tensor component in the time direction,[33]

$$\frac{F''}{F} + 2\frac{F'G'}{FG} = 0.$$

---

[33] A. L. Besse, *Einstein Manifolds* (Springer, Heidelberg, 1987) Ch 3.

However, we would never have been aware of the fact that K_{02} is related to the tidal forces by way of (2.96). This again can be chalked up to a 'comedy of coincidences.'

To recover the usual form of the Schwarzschild metric, we set: $G(r) = \tilde{r}$. The metric, (2.85), then becomes:

$$c^2 d\tau^2 = (1 - 2\mu/\tilde{r})c^2 dt^2 - \frac{d\tilde{r}^2}{1 - 2\mu/\tilde{r}} - \tilde{r}^2 d\sigma^2, \qquad (2.100)$$

which is the usual form of the Schwarzschild metric. But coordinate transforms cannot be applied to (2.100) that would eliminate the 'inessential' singularity at $\tilde{r} = 2\mu$, without changing the meaning of the metric entirely.

The tidal forces scale as the square root of the Kretschmann constant: ,

$$R_{\alpha\beta\mu\nu} R^{\alpha\beta\mu\nu} = 48\mu^2/r^6.$$

The Kretschmann constant is commonly used to assert that the putative singularity at 2μ is removal, but the essential singularity at 0 is not! Anyone who has been familiarized with non-Euclidean metrics would know that 2μ is the boundary of an impenetrable disc, just like the square root of the inverse density is the radius of the hyperbolic disc in the Schwarzschild inner solution.

The hyperbolic universe terminates on the rim of the disc, and there would be no meaning to prolong it further. Furthermore, it would take an inhabitant an infinite amount of time to reach the rim since his clocks and rules would shrink along with him. This can be likened to Zeno's paradox of Achilles and the Tortoise.

The non-vanishing of the Kretschmann curvature is, indeed, troublesome when the Ricci components and the scalar curvature are all zero. The tidal forces are all non-zero, but, rather, conspire to average to zero.

If we contract the Einstein equations with a time-like unit vector, we get the scalar curvature of the spatial directions proportional to the energy density. But, how can this be when the Schwarzschild solution arises from Einstein's condition of emptiness implying that the energy-stress tensor is identically zero? In other words, how can the scalar curvature of the spatial dimensions be non-vanishing at the same time the energy density vanishes implying a vanishing spatial curvature?

And, furthermore, why should the averaging of the sectional curvatures be any indication of geodesic deviation when there is no meaning to the averaging of the components of the tidal forces? What is there so special in the average?

Does it make a difference whether tidal forces scale as μ/G^3, or $G\rho$? Not when it comes to the Roche limit of a planet. This is the minimum distance that a body can be from the planet and still be held to it by gravitational forces only. In this limit, the rotation about the planet has no bearing.

An example is given by Neptune's rings that do not depend on the orbital speeds of the constituent particles in the ring. Neither should angular momentum have an effect on the Schwarzschild metric; in fact, gravitational forces do not act on body orbiting the central body because there is no body. This is due to the classical Newtonian characteristic that the properties of orbiting bodies do not enter into the equations. The mass of the peripheral body appears on both sides of the equation, once as Newton's second law, and the other by his inverse square law, and so the peripheral mass cancels out of the equation.

2.11.3 The Friedmann Metric

It was Friedmann's intention to create "a non-stationary world with constant negative spatial curvature and with a positive matter density.[34] However, he missed his mark when he introduced a time-dependent scaling factor that represented the radius of the universe.

Friedmann began with Poincaré's generalization of his half-plane model by moving off the plane into space. In the same way the real axis is 'the boundary at infinity' of the half-plane, with line element:

$$R(t)\frac{\sqrt{dx_1^2 + dx_2^2}}{x_2},$$

in the (x_1, x_2)-plane, so the boundary at infinity of the half (x_1, x_2, x_3)-space has a

---

[34] A. Friedmann, "On the possibility of a world with constant negative curvature", translated in *Cosmoogical Constants*, J Bernstein & G Feinberg eds (Columbia UP New York, 1986), p.59.

line element: [35]

$$R(t) \frac{\sqrt{x_1^2 + x_2^2 + x_3^2}}{x_3}$$

Both have a *time-dependent* "spatial curvature of our world [which] is negative and proportional to $-1/R^2(t)$."

As time goes on, its natural that things get generalized, but, more often, they become twisted, and the Friedmann metric is no exception. If we use q rather than r as the radial coordinate, the Friedmann metric mutated into what was to be known as the Friedmann-Walker-Robertson (FWR) metric

$$ds^2 = -dt^2 + R^2(t) \left[\frac{dq^2}{1 - kq^2} + q^2 d\sigma^2 \right], \tag{2.101}$$

where k is a constant. This constant, miraculously, distinguishes between no-less than three types of universes: 'open' $k > 0$, 'closed' $k < 0$, and 'flat' $k = 0$, universes.

However, this does not concur with the time-dependent, negative curvature, $-1/R(t)^2$, of Friedmann's model. Friedmann explicitly used the half-space model of Poincaré to ensure negative, but time-independent curvature. The transition between the Friedmann and FWR models camouflages this entirely.

We begin with a spatial metric of the form

$$ds^2 = dr^2 + f^2(r, t) d\sigma^2,$$

where $f(r, t)$ can be both a function the radial coordinate as well as time. Now, a preposterous distinction is made between 'floating', r, and fixed observers, $\bar{r}$ such that $r = R(t)\bar{r}$. If this is true, then the space component of the metric should also pick up a time dependency of the form $\bar{r}^2 dR^2(t)$, which it doesn't.

The metric then takes the form:

$$ds^2 = R^2(t) \left[d\bar{r}^2 + q^2(\bar{r}) d\sigma^2 \right],$$

---

[35] J Stillwell, *Sources of Hyperbolic Geometry* (Am Math Soc, Providence R I, 1996) p 38.

where the *product* $R(t)q(\bar{r})$ has been substituted for the function, $f(r,t)$. But, this function is a function of $\bar{r}$ and not r so it should read $f(r,t) = R(t)q(R(t)\bar{r})$, and q is a function of t also! Why should the coefficient of the radial component of the metric be independent of r?

All this is necessary to keep the time-independent, spatial curvature k time-independent. This is borne out by calculating the spatial sectional curvature,

$$\frac{k - \dot{R}^2}{R^2},$$

which is **not** time-independent, as Friedmann would have it. However, there is no sectional curvature of the sort for values of k other and 1. Thus, nothing has been gained. This, nevertheless, should not dissuade us from following the path of contorted reasoning.

We have no reason to distinguish between q and R as being functions only of r and t respectively. Rather than following the flock [36] we consider the scale factor to be both a function of r and t instead of writing it as a product for, at face value, there would be no advantage.

Thus, the metric is:

$$ds^2 = -dt^2 + R^2(r,t)\left\{dr^2 + r^2 d\sigma^2\right\}. \tag{2.102}$$

The diagonal components of the Ricci tensor are:

$$R_{tt} = -3\frac{\ddot{R}}{R},$$

$$R_{rr} = R\ddot{R} + 2\dot{R}^2 - 2\frac{R''}{R}, \tag{2.103}$$

and

$$R_{\theta\theta} = 2q^2\dot{R}^2 + q^2 R\ddot{R} - R'^2 - RR'' + 1.$$

---

[36]T A Moore, *A General Relativity Workbook* (2010), pp 265ff.

Parenthetically, we note that the linearization of (2.103), when set equal to zero, becomes the sourceless GW equation. Had Riemann heard of (or appreciated) an indefinite metric, he could have predicted GWs, over a half a century before Einstein by linearizing the Ricci tensor and setting it equal to zero.

Modelling the universe as a 'perfect' fluid, the energy-stress tensor is:

$$T_\nu^\mu = (\rho(t) + p(t))u^\mu u_\nu + p\delta_\nu^\mu,$$

where the density, ρ, and pressure, p, are functions of time only, and u_ν are the 4-velocities, $u^\nu u_\nu = -1$, remembering that we are dealing with geodesic motion.

This hardly can be considered as a 'perfect' fluid. If we write the Einstein field equation as:

$$R_{\nu\mu} = 8\pi G \left(T_{\nu\mu} - \frac{1}{2}g_{\nu\mu}T \right),$$

where $T = -(\rho - 3p)$, it follows that

$$R_r^r - R_\theta^\theta = -\frac{1}{R^2} \left(\frac{q''}{q} + \frac{1 - q'^2}{q^2} \right) = 0, \qquad (2.104)$$

since the energy-stress components for the pressure are all equal.

Condition (2.104) is rather interesting since it equates two sectional curvatures,

$$-q''/q, \quad \text{and} \quad \frac{1 - q'^2}{q^2}.$$

It is important to note that there is no time-independent, scalar curvature, k, in the second sectional curvature. That arises from a constant of integration.

The simplest assumption is to assume that:

$$\frac{q''}{q} = K, \qquad (2.105)$$

which is the Jacobi equation. If we went ahead and integrated (2.104), we would have found k as an arbitrary constant of integration. But, since the integration is with respect to r, and q must be considered also a function of t, the constant K must also be considered a function of time — the variable held constant in the integration.

Recalling our earlier discussion surrounding (2.81), all we can deduce from the Jacobi equation, (2.105), is that the rate of spreading depends on Gaussian curvature. For $K > 0$, the radial geodesics spread faster than they do in Euclidean geometry, while for $K < 0$, they spread less rapidly like those on a sphere.

And since it is arbitrary, and a sole function of time, the expressions for q,

$$q(r,t) = (1/\sqrt{K})\sinh(\sqrt{K}r),\qquad(2.106)$$

covers all three geometries: hyperbolic $K > 0$, elliptic $K < 0$, and Euclidean, $K = 0$.

Confusion abounds when attempting to distinguish between *two* curvatures: a 4-dimensional curvature, K, and a 3-dimensional curvature, k [37]. Robertson writes the constant curvature equations as:

$$\ddot{R} = KR, \qquad \text{and} \qquad \dot{R}^2 + k - KR^2 = 0.$$

Notice that time and not space is involved. Robertson then goes on to list all the different possibilities, and give credence to them. Why time and not space is involved is that after integrating,

$$R'' - 1 = KR^2,$$

Robertson then writes the metric in the form (2.101).

Now, once in the form of the metric (2.101), there are only two non-zero sectional curvatures: a temporal one,

$$-\frac{\ddot{R}}{R},\qquad(2.107)$$

and a spatial one,

$$\frac{K - \dot{R}^2}{R^2}.\qquad(2.108)$$

The latter, (2.108), makes absolutely no sense, since it is a curvature of a curvature. This goes back to our previous remark that K must be a function of time, according the conventional derivation.

---

[37] H P Robertson & T W Noonan, *Relativity and Cosmology* (W B Saunders, Philadelphia, 1968) p 346.

This illustrates a salient feature of GR: you can't transform one metric into another and expect that same physical conditions to apply.

It is commonly believed that the Schwarzschild singularity is an inessential, or 'removable,' singularity. As we have mentioned, that is tantamount to removing the periphery of the disc in the Poincaré disc model of hyperbolic plane. We will give another example where a coordinate transformation changes the model entirely.

The vanishing of the difference of the sectional curvatures leading to (2.106), makes one think that their sum should be related to a force. However, the Einstein tensor,

$$G_{\mu\nu} \equiv R_{\mu\nu} - \frac{1}{2}Rg_{\mu\nu}, \qquad (2.109)$$

contains an appendix $Rg_{\mu\nu}/2$, that contains the scalar curvature, R, which is required if the divergence of the Einstein tensor, G, is to vanish on account of the Bianchi identities . Einstein associated the vanishing with conservation conditions. But, conservation of what? when energy and momentum cannot be localized.

On the right-hand side of (2.103) we have the inverse square of the Newtonian free-fall time. Classically, the force is proportional to $\dot{R}\dot{R}'$ which would indicate a Hookean law:

$$F = R/\tau_N^2,$$

where $\tau_N = 1/\sqrt{(G\rho)}$, where $\rho \sim M/R^3$, independent of K. It is, indeed, strange that this force should be operative no matter what scenario, open, closed or flat, is chosen for the evolution of the universe.

At zero pressure, the R_{rr} component of the Ricci tensor vanishes:

$$\frac{\dot{R}^2}{R^2} - 2\frac{\ddot{R}^2}{R^2} - \frac{K}{R^2} = 0, \qquad (2.110)$$

without what Friedmann refers to as the 'Gleide member,' or the notorious cosmological constant. Integrating (2.110) in time gives:

$$\dot{R}^2/c^2 = \frac{A}{R} + 1,$$

where A is an arbitrary constant of integration. Like all other *arbitrary* constants of integration in GR, it is given a specific meaning.

In this instance, it is used in the definition of the mass density as:

$$\rho = \frac{3A}{4\pi G R^3},$$
(2.111)

so if $A > 0$, so too, will be ρ. Consequently, the expression for the density is:

$$\frac{8\pi G}{3}\rho = -\frac{1 - \dot{R}^2}{R^2},$$
(2.112)

and this is what is obtained from the time component of the Einstein field equations. However, Einstein's equations do not specify that the density is defined by (2.111), or, equivalently, $\rho R^3 = $ const. So let us go back to see what happens when the pressure is non-zero.

The field equations give the pressure as:

$$\frac{8\pi G}{c^2}p = -2\frac{\ddot{R}}{R} + \frac{1 - \dot{R}^2}{R^2}.$$
(2.113)

Multiplying both sides by $\dot{R}$, and integrating in time give:

$$\frac{8\pi G}{c^2}\int pR^2 dR = -\int (2\dot{R}\ddot{R} + R^2 - K)\dot{R}dt.$$

Since p is independent of R, integration and use of (2.112) result in:

$$pR^3 = -\rho c^2 R^3,$$
(2.114)

and not in the usual quoted form:

$$\frac{d}{dt}(\rho c^2 R^3) - p\frac{dR^3}{dt} = 0.$$
(2.115)

Equation (2.115) evokes a false analogy with the first law of thermodynamics. It would imply that the universe is adiabatic, which is really not such a far-fetched idea.

Moreover, if we introduce (2.114) into (2.115) we get the nonsensical condition that $\sqrt{\rho}R^3 = $ const., instead of (2.111). Relation (2.114) is even more telling.

Robertson gets the same expression as (2.114), except he claims that it holds for "constant Gaussian curvature." We have derived it without any supposition on the value of K, except, of course that it be constant, which everyone agrees upon. It shouldn't be forgotten that it is really a function of time, the variable held constant during the integration.

Robertson concludes: [38]

> Thus all cosmological models will admit the maximal group of automorphisms. . . require a negative pressure equal to the energy density! Under the interpretation of cosmological pressure given above, there exists no physical solution to eq. [(2.114)] because the only possible negative contribution to the pressure comes from the gravitational energies of the galaxies. Equation [(2.114)] therefore requires that both p and ρ vanish. But then we may ask what it is, if not matter, that follows the fundamental world-lines. The constant-curvature models can be rescued by relaxing either the field equations of general relativity or the hydrodynamic interpretation of cosmological pressure.

In fact, both must be relinquished. This should have been a non-starter from the very beginning, so why did he pursue the matter further?

It is even more astonishing how Guth [39] could have used the FWR metric to obtain cosmic expansion which he called inflation by requiring an enormous increase in entropy in an adiabatic universe! Perhaps he could be forgiven because his roots were particle physics, but not the tremendous support he received from cosmologists who supported his idea. Nothing is spared when it comes to saving the standard model.

If there is pressure, and not the effect of pressure by the action of gravity— which in

---

[38] *op. cit.* p 374
[39] A H Guth, *The Inflationary Universe* (Addison-Wesley, Reading MA, 1997).

any event should act in the opposing sense — it should definitely appear in the energy conservation equation, which (2.112) was interpreted as.

A closely related model is the de Sitter universe. It begins with the Schwarzschild inner metric:

$$ds^2 = -\left(1 - \frac{r^2}{R^2}\right) dt^2 + \frac{dr^2}{1 - r^2/R^2} + r^2 d\sigma^2, \tag{2.116}$$

where σ^2 is a 2-sphere.

The metric (2.116) is conformal with respect to the Beltrami metric, which describes the Poincaré disc model. The constant R is the rim of the disc, the boundary of the hyperbolic plane.

Now look what a transformation to new coordinates, $r', \vartheta', \varphi', t'$, does. The transformation is:

$$r' = \frac{r}{\sqrt{1 - r^2/R^2}} e^{-t/R}, \qquad\qquad t' + \frac{R}{2} \ln\left(1 - \frac{r^2}{R^2}\right), \tag{2.117}$$

$\vartheta' = \vartheta$, and $\varphi' = \varphi$. As was shown independently by Lemaitre and Robertson, this transformation brings the metric into the form:

$$ds^2 = -dt^2 + e^{2ct'/R}\left(dr^2 + r'^2 d\sigma'^2\right) \tag{2.118}$$

where $\sigma' = \sigma$. Gone is the boundary of the hyperbolic plane, and in order to make contact with the metric (2.102) all that is necessary is to set $R(t) = e^{2t'/R}$. So a mere, innocuous transformation as (2.117) can raise havoc in the model, transforming into something entirely different from what it started out to describe.

Just look at the predictions it makes. In terms of the original coordinate system, it is found that:

$$r^2 = \frac{r'^2}{e^{-2t/R} + r'^2/R^2},$$

where r' is constant for each nebula. From this it is concluded that: [40]

---

[40] Moller, *op. cit.* p 368.

> the nebulae according to the Weyl hypothesis are freely falling particles
> which start at the point $r = 0$ at $t = -\infty$ and end at the antipodal point
> $r = R$ for $t \to \infty$.

This is a simple statement that it takes an infinite amount of time for the Poincarite, the inhabitant of the disc, to reach the rim.

The exponential shift of the wavelength of light is also a consequence of the transformation (2.117). What was not in the original model was written in by hand, and it can hardly be considered as "an explanation of the actual red shift observed by Hubble and Humason."

The manner in which de Sitter obtained his (or more correctly Schwarzshild) solution by integrating the Einstein field equations for a perfect fluid required the equation of state:

$$\rho c^2 + p = 0$$

This can hardly be considered as an equation of state because for positive rest mass density the pressure "is negative and very large. . . it would be quite incompatible with any known material." [41] Moreover, the conclusion that:

> the de Sitter model corresponds to an empty universe containing no
> appreciable amount of matter and radiation, the stars and nebulae being
> treated as a kind of test bodies which do not contribute essentially to the
> gravitational field.

Moller then tries to invoke Mach's principle which is not incorporated into GR, to the chagrin of Einstein.

Then what kind of perfect fluid are we talking about? The gravitational field is present because of the form of the metric, (2.116). And if the stars and nebulae don't contribute to it, what contributes to it? And finally, the large negative pressure that was rejected as being unphysical forms the driving force for the inflationary scenario. How worse a situation can there be?

---

[41] de Sitter, *op. cit.* p 369.

In the early days of what has become known as the standard model, Gamow rewrote (2.112) as an energy conservation equation:

$$v = [2(\mu/R - E)]^{1/2},$$

where $\mu = GM$, the gravitational parameter, and the three scenarios for k are related to the sign of the total energy, E. But, because $k = k(t)$, an arbitrary function of time, Gamow is admitting that energy is not conserved. Since he was oblivious to this fact, he plowed ahead and distinguished between elliptic and hyperbolic geometries for which

> imaginary and real values of R correspond to an unlimited expansion (in the case of super-escape velocity), and to the expansion which will ultimately [be] turned into a contraction by the force of gravity (sub-escape) velocity.

However, the energy conservation equation for ρ is independent of the pressure whether it be zero or non-zero. It does appear in what everyone rushes to identify as the adiabatic condition of the second law, (2.115). Yet, we know from Euler's equation for a perfect fluid that the pressure does appear in the equation of motion, which can be written as

$$\dot{R}\ddot{R} = -\frac{1}{\rho} - \frac{M}{R^2}\dot{R}. \tag{2.119}$$

If $p = 0$, $\rho R^3 =$ const., so the total mass, M, is constant. But, in order to go from (2.119) to (2.112), we are constrained to consider the mass constant during the integration in time, and the pressure does not appear anywhere. Why then does it appear in the 'first' law, (2.115), and doesn't appear in the mechanical conservation of energy, (2.112)?

Notwithstanding this 'minor' blemish, Gamow [42] concludes that:

> at the epoch when the mean density of the universe was of the order of 10^6 g/cm^3, the expansion must have been proceeding at such a high rate that this high density was reduced by an order of magnitude in only about one second.

---

[42]G Gamow, "Expanding universe and the origin of the elements," *Phys Rev* **70** (1946) 572.

Folklore has it that the first equation gives the condition for the critical mass, in terms of the Hubble 'constant', $H \equiv \dot{R}/R$, as:

$$\rho_c = \frac{3H^2}{8\pi G},$$ (2.120)

and

$$\rho - \rho_c = \frac{3K}{8\pi G},$$

which vanishes when $K(t) = 0$. But this must apply to a specific instant in time, and not just any time. And how does the pressure influence this critical density? How then can there be the claim that over 85% of the mass is missing from the universe when its critical value, ρ_c, has no meaning?

2.11.4 The FWR Metric

Again our comedy of coincidences intervenes, and the Friedmann equation for the mass density boasts that it can be derived solely from classical physics, so it can't be wrong.

Consider the universe at two different times. A universe in steady motion with only gravitational energy will satisfy the balance equation:

$$\frac{1}{2}v^2 - \frac{GM}{R} = \frac{1}{2}v_0^2 - \frac{GM}{R_0},$$

where the subscript '0' refers to here and now, t_0, and the unsubscripted symbols refer to some time in the future. If we introduce the present time Hubble law, $v_0 = HR_0$, and write the total mass of the universe in terms of its density, $\rho_0 = M_0/4\pi R_0^3/3$, and then letting $t \to \infty$, together with $R \to \infty$, we will come out with:

$$v^2 \to R_0^2 \left(H_0^2 - \frac{8\pi G}{3}\rho_0 \right).$$ (2.121)

If it should happen that at some point in the evolution of the universe, its speed should vanish, then saith the soothsayers a critical density, (2.120), should ensue.

According to many, if not all, authors of undergraduate texts in cosmology, this is "amazing enough [since] a detailed analysis of the theory of general relativity leads to

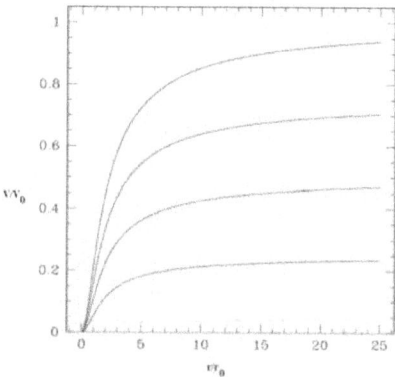

Figure 2.28: Flat rotation curves of spiral galaxies as a function of their distance from the center.

the same constraints and values for ρ_c as the classical description." [43] If anything, this should warn us that something is definitely amiss.

It's one thing to let parameter, K, take on different integers, even though it is a function of time, and, yet, another to assume that $v^2 < 0$ represents a 'big crunch.' It is even more astonishing that the criterion for the critical density was derived under the supposition that $R \to \infty$ as $t \to \infty$. Isn't there any check for internal consistency?

2.12 Flat rotation curves of galaxies

It is a well-known fact that the rotation curves of galaxies are asymptotically flat, as shown in Fig (2.28):

Luminosity and line-widths are the standard bearers of observational astronomy. Undoubtedly, the best known, and most widely used, extragalactic empirical relation between a galaxy's luminosity and its 21-centimeter hydrogen line width, is the Tully-

---

[43]H S Goldberg & M D Scadron, *Physics of Stellar Evolution and Cosmology* (Gordon & Breach, New York, 1987) p 318.

Fisher relation. It has also been used as a distance indicator, although its physical foundations have remained obscure.

In most bright galaxies, stars account for the majority of baryonic mass so luminosity is a good measure of their mass. In more slowly rotating galaxies, which are predominantly gaseous, the luminous Tully-Fisher relation no longer holds, but an analogous relation appears to hold if luminosity is replaced by baryonic mass. The power law relation between baryonic mass, M, and rotational velocity, V, was to be expressed as the logarithmic relation:

$$\ln M = a \ln V - \ln b, \tag{2.122}$$

where M is the sum of the mass of observed stars and gases, and a and b are constants.

Empirically, the Tully-Fisher law sets the velocity exponent $a = 4$, and the intercept was found to be Ga_0, where $a_0 \simeq 10^{-10} m/s^2$ appears to be a universal galactic acceleration constant. Without knowing its value, Milgrom likened a_0 to Planck's or Boltzmann's constant, since in the limit $a_0 \to 0$, Newtonian dynamics should apply. However, in a regime of small accelerations, like the typical centripetal acceleration in a galaxy which is 11 orders of magnitude less than the surface gravity on Earth, a modified gravitational acceleration would apply that would be the geometric mean of Newtonian gravitational acceleration and a_0.

Milgrom's expression [44] for the gravitational acceleration,

$$g = \sqrt{g_N a_0}, \tag{2.123}$$

is the geometric mean of the Newtonian acceleration, g_N, and his universal galactic acceleration constant, a_0. Introducing the centripetal acceleration, v_{asy}^2/r for g, (2.123) becomes:

$$v_{asy} = \sqrt[4]{GMa_0}, \tag{2.124}$$

which is entirely independent of the radius, if the total mass M is constant. The asymptotic velocity depends solely on the fourth root of the mass of the central object.

Milgrom's law reproduces the flatness of the galaxy rotation curves, and the slope of the Tully-Fisher relation with baryonic mass replacing luminosity.

---

[44]M Milgrom, "A modification of the Newtonian dynamics as a possible alternative to the hidden mass hypothesis," *Astrophys J* **270** (1983) 365-370.

Another approach that takes into account the distinction between luminosity and mass modifies Newton's law of gravitation instead of the law of inertia, viz., [45]

$$g(r) = -GM \left(\frac{1}{r^2} + \frac{1}{r_0 r} \right),$$

(2.125)

where r_0 has dimensions of a length comparable to a typical galactic radius. The result is similar: For distances much smaller than r_0, Newton's inverse square law goes over into an inverse *linear* law.

It is found that the mass to luminosity ratio is in the same proportion as the ratio r/r_0 and both increase linearly as a function of scale size. The data indicates a value of r_0 of a few thousand light-years.

Although empirical evidence lends support to both (2.125) and (2.123), over times scales of the order of the free-fall time, $\sqrt{r_0^3/GM}$, which is of the order of thousands of millions of years on the galactic scale, we would like to understand the latter in theoretical terms.

2.13 Attempts to Modify Newtonian Dynamics

Newton's law of gravity has already presented paradoxical results that were known to Newton himself. Newton showed that a spherical body exerts gravitational attraction as if its mass were concentrated at the center of the body.

Yet, if the universe is uniformly populated with matter enclosed in a set of concentric shells, the stars on one side of a shell would exert an equal and opposite attraction to those on the opposite side of the shell. Hence, the stars in any given shell would exert no net force on the matter which is enclosed in it.

Among other things, this goes against the grain of Mach's principle whereby the inertia of matter is determined by the distribution and masses of all other bodies in the

[45] A E Wright, M J Disney & R C Thomson, "Universal gravity: was Newton right?" *Proc ASA* **8** (1990) 334-338.

universe. Actually, Mach's principle is an echoing of what Bishop Berkeley proposed
a century earlier.

Moreover, the number of stars in any shell is proportional to the surface area, or the
square of the radius. And since each star would exert an attraction proportional to
the inverse square of the radius, the intensity would be independent of the radius of
any shell. As we sum up all the concentric shells, the opposing forces get stronger and
stronger, and at an infinite distance become infinitely great.

There must be something wrong with the exact isotropy of Newton's universe which
would also lead to Olbers's paradox. If the universe were static and infinite, the night
sky would be set ablaze with light, yet the night sky is dark.

Forces of a higher power in the inverse radius, greater than second, would mean that
the gravitational force would decrease faster than Newton's law. It would leave open
the possibility of infinite universes. Alternatively, a law of universal gravitation that
tends to an inverse linear law on large scale would mean that the force would increase
faster than Newton's law so that there could be no 'open' universes. A finite universe
would also impose that the gravitational force be finite. Gravity would win in the
end, and the final act would be a contraction to a singularity. It would also revive
Berkeley and Mach's principle where matter in the distant universe would appear more
important locally than previously imagined.

Parenthetically, we mention that the finiteness of the gravitational force was already
considered by Laplace. [46] Laplace's theory can be seen as an echoing of Le Sage's particle theory of gravity [47] whereby very high speed gravitational particles would be
absorbed by matter.

Accordingly, Laplace assumed that the gravitational constant, G, was not an absolute
constant, but, rather, varied exponentially with distance since gravity can be 'absorbed'

---

[46]P S Laplace, "Traite de Mecanique Celeste," in *Oeuvres de Laplace*, (Gauthier-Villar, Paris, 1880) vol 5, book 16, chapter
4.
[47]*Pushing Gravity*, M R Edwards, ed. (Apeiron, Montreal, 2002).

by matter. His gravitational potential reads:

$$\Phi = \frac{G_N M}{r} e^{-\gamma r},$$

where G_N is the Newtonian gravitational constant, and γ represents the attenuation of the intensity of gravity. Laplace placed an upper bound on the attenuation coefficient, $\gamma \leq 7 \times 10^{-18}\ m^{-1}$, for, otherwise, the effects of his modification would have observable effects on the solar system.

Laplace's force:

$$F = -G_N M e^{-\gamma r} \left(\frac{1}{r^2} + \frac{\gamma}{r} \right),$$

would be analogous to (2.125) with the attenuation constant, $\gamma = 1/r_0$, and a modified gravitational constant, $G_N e^{-\gamma r}$.

2.14 Light deflection in two and three dimensions

General relativity deals with an indefinite metric whose vanishing determines the null geodesics along which light rays propagate. In non-Minkowski (curved) space-time, light rays will not propagate at the speed of light, c, because they are slowed down by the presence of the gravitational field, Φ. The gravitational field modifies the space and acts like an index of refraction.

This, in essence, constitutes Einstein's original derivation [48] of the angle of deviation that a light ray succumbs to if it propagates in a medium whose index of refraction is:

$$n = 1 + \Phi/c^2, \tag{2.126}$$

where $n = c/c_0$, with c_0 the speed of light in vacuum.

---

[48] A Einstein, "On the influence of gravitation on the propagation of light," *Ann d Phys* **35** (1911).

There are at least four inaccuracies in Einstein's derivation. First, he derives it from a classical Doppler shift in the frequency. A clock will go more slowly in a medium with an index of refraction (2.126) than one in vacuum by an amount:

$$\nu_0 \left(1 + \frac{\Phi}{c^2}\right),$$

where ν_0 is the frequency of emission. Since he is considering that light passes through a medium formed by a gravitational potential, it is the wavelength that changes with the speed of light, and not the frequency. In other words, it is Snell's law, and not Doppler's, that applies.

Second, it is not a modification of the "principle of the constancy of the velocity of light [which] holds good in a different form from that which usually underlies the ordinary theory of relativity." Consequently, the speed of light in a gravitational field will be given by:

$$c_0 \left(1 + \frac{\Phi}{c^2}\right),$$

where c_0 is the speed of light in vacuum.

Third, Einstein appeals to Huygens's principle rather than to Snell's law. Huygens's principle states that everywhere a luminous disturbance touches it, it determines the form of a new wave. That is, each point on the wave front emits a semicircular wavelet that propagates at a constant velocity. However, it is impossible to predict the velocity of the wave front crossing the boundary by Huygens's principle; for that you need to know the ratio of the two velocities and that is given by Snell's law.

Einstein appeals to Huygens's principle to masquerade the fact that he needs to reduce his $3D$ theory to a $2D$ one. This is the fourth inaccuracy in his derivation. For he expresses the difference in the speed of light as $-\partial c/\partial n$ for a ray which is "bent toward the side of increasing n." Einstein then *defines* the deflection angle α as an *integral* along the undeflected path s:

$$\alpha_E = -\frac{1}{c^2} \int_{-\pi/2}^{\pi/2} \frac{\partial \Phi}{\partial n} ds = \frac{1}{c^2} \int_{\theta=-\pi/2}^{\theta=\pi/2} \frac{GM}{r^2} \cos\theta ds = 2\frac{GM}{c^2\Delta}, \qquad (2.127)$$

where "Δ is the distance of the ray from the center of the body." This he then equates with the radius of the sun, and in so doing comes out with 0.83 seconds of an arc. To compute the integral, $r d\theta$ must be written in place of ds, and r must be associated with the impact parameter, Δ.

Here, Einstein is implicitly appealing to Newton's shell theorem which says that a gravitational field outside a spherical shell having a total mass M is the same as if the entire mass is concentrated at the center, or if non-spherical, at the center of mass. This law, however, does not hold for the sun, since it is rotating, and angular momentum is not conserved. So Newton's shell theorem does not apply to the sun, and Einstein could not have identified the radius of the sun with "the distance of the ray from the center of the body."

It is also curious why no one has ever contested the necessity of integrating over the unperturbed path of the ray in (2.127), especially when the value of the deflection angle is extremely sensitive to the limits of integration.

In contrast, Soldner's original treatment, which he derived eleven decades earlier, [49] needed no integration along the path of the ray. Soldner simply used the conservation of energy:

$$E = \frac{1}{2}\left(\dot{r}^2 + r^2\dot{\phi}^2\right) - \frac{GM}{r} = E = \frac{1}{2}c^2,$$

and the conservation of angular momentum, $r^2\dot{\phi} = \text{const}$, to derive the same deflection angle. Since the total energy is positive, $c^2/2$, the orbit is a hyperbola. The angle of intersection of the two asymptotes, $Y = \pm bx/a$, is $\alpha = 2\tan^{-1}(b/a)$, which, upon inversion, gives:

$$\tan(\alpha/2) = \frac{a}{b} = \frac{GM}{2E} \cdot \frac{1}{b},$$

where a, and b, are the lengths of the transverse, and conjugate, semi-axes. Since $GM/c^2 \ll b$, where b is the distance of the asymptotes from the foci, or the impact parameter, the deflection angle can be written as:

$$\alpha = \frac{2GM}{c^2} \cdot \frac{1}{b},$$

which is identical to Einstein's expression when b is set equal to Δ, the impact parameter. But, there is no necessity to integrate along the path of the ray!

---

[49]H-J Treder, "On Soldner's value of the Newtonian deflection of light," *Astron Nachr* **302** (1981) 275-277, where Soldner's article is referenced.

And since there is no torque in $2D$, it resolves the problem of the sun's non-conservation of momentum in $3D$. Moreover, because it is a gas, the sun spins faster at the equator than at the poles. In fact, both Kant and Laplace, when they put forth their nebular theory of the formation of the solar system, failed to consider how the infant sun could lose angular momentum. These considerations do not appear in $2D$. However, even more important, is that no integration is needed in Soldner's case which calculates simply the angle of the asymptotes of the hyperbolic orbit.

In the section on dual laws, we showed that for every power law of attraction there is a unique dual law. For instance the dual law to Hooke's law was seen to be Newton's law of gravitation. This can now be seen as consequence of Bertrand's theorem which states that:

> Among the central force laws with bounded orbits, there are only two types of central force potentials for which all *bound* orbits are also *closed* orbits. They are the Hookean force and the Newtonian inverse square law.

Orbits in $2D$ are no longer ellipses, although they are *bound* orbits, they are not *closed* orbits. An example is a radial orbit where two objects move toward, or away from, each other. The angular momentum is zero, and the eccentricity is unity.

According to our dual law criterion, (2.73), the law of gravitation in $2D$ would be its own dual, $a = \tilde{a} = -1$, and the complex plane transform would be $z = 1/w$.

Attempts to generalize the inverse square law were based on Gauss's law which states that

> the surface integral of the gravitational field $\vec{g}$ over a closed surface $\partial \vec{S}$ is proportional to the amount of mass, M, that it encloses,

$$\int_{\partial S} \vec{g} \cdot d\vec{S} = -4\pi GM.$$

In $3D$, a sphere with a surface area $4\pi r^2 \hat{r}$ that encloses a (point) mass M is $4\pi r^2 \hat{r} \cdot \vec{g} = -4\pi GM$, where appeal has been made to Newton's shell theorem for reducing the

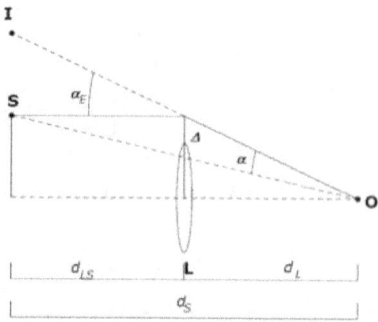

Figure 2.29: The source S, the lens, L, and the observer, O, all lie on the same line. The angular
distance α the the observer measures is related to Einstein's deflection α_E at the
lens by the ratio of the distance from the source to the lens, d_{SL} and the distance
from the observer to the source, d_S.

total mass to a point.

Alternatively, in $2D$, the closed surface is reduced to the circumference of a circle so
that Gauss's law would read, $2\pi r \hat{r} \cdot \vec{g} = -4\pi GM$; the corresponding force law is:

$$g = -\frac{2GM}{r},$$

This new force law implies a logarithmic potential. And although the orbits are bound,
they are not closed. Quite surprisingly, the reduction in dimension converts a closed
orbit into an open one.

Einstein interprets his angular correction as "angular distance of the star from the
center of the sun [which] appears to be increased by that amount." The deflection of a
lens L of a source S with respect to an observer O is shown in Fig (2.29).

The Einstein angle of deflection, α_E, is related the angular distance α by:

$$\alpha_E = \alpha \cdot \frac{d_S}{d_{LS}}$$

which is the ratio of the distance from the observer to the source and the distance from the lens to the source.

Einstein's angle can be calculated from the derivative of the $2D$ potential,

$$\psi = \frac{4GM}{d_L} \ln \alpha, \tag{2.128}$$

with respect to the angular radius, α, viz.,

$$\alpha_E = \frac{1}{c^2} \cdot \frac{d\psi}{d\alpha} = \frac{4GM}{c^2 \Delta}. \tag{2.129}$$

The impact parameter, Δ, is the physical radius of the sun provided all the mass can be concentrated at its center. Consequently, the angle of deflection is:

$$\alpha = \frac{4GM}{c^2 \Delta} \cdot \frac{d_{SL}}{d_S}. \tag{2.130}$$

The commonly accepted derivation of the lensing equation makes use of the so-called Fermat potential formalism. [50] There one tries to marry a $4D$ metric, like the Schwarzschild metric, with the Shapiro time delay, which is $2D$. Since the $4D$ metric contains a $3D$ potential, an integration over the $3D$ potential must be performed-without specifying the limits of integration-in order to obtain the Shapiro delay. The variational principle appears as the difference between geometrical and temporal factors. This is so in order to obtain the deflection angle, but, in reality, no such decomposition exists.

Not only does the gradient of the logarithmic potential give the angle of deflection–without any integration over an unperturbed ray–it is also of interest in the study of elliptical galaxies and galactic haloes.

---

[50]K Kuijken, "The basics of lensing," in *Gravitational Lensing: a unique tool for cosmology* D Valls-Gabaud & J-P Kneib, eds *ASP Conference Series*, (2002).

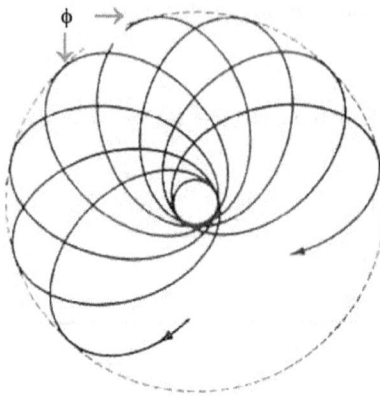

Figure 2.30: The angle formed from two consecutive apsides in ϕ.

2.15 Newton's theorem on revolving orbits

Newton's theorem of revolving orbits answers the question of what type of central force is necessary so that its radial motion remains the same while the angular speed increases by a factor m. Newton used his theorem to study the rotation of orbits, referred to as aspidal precession, that were known for the moon and the planets.

Newton addressed the problem of rotating orbits in Propositions 43-45 of Book I in his first edition of *Principia*. Specifically, in Proposition 43 he identifies what force is necessary as a central force in order to keep the radial motion the same while varying the angular speed. In Proposition 44, Newton shows that the additional force is an inverse cube of the separation, and in Proposition 45 he extends his result to arbitrary central forces, for orbits of small eccentricity. All this became a criterion for establishing his inverse-square law.

Newton derives a relation between the central force exponent, n, in μr^{n-3} and the apsidal angle, ϕ, the angle at the force center between highest and lowest apses, or the apocenter and pericenter respectively, as as illustrated in Fig (2.30).

Newton gives the apsidal angles as:

$$\phi = \frac{\pi}{\sqrt{n}} \tag{2.131}$$

By the manner in which Newton derives (2.131), it is restricted to only those orbits 'approaching very near to circles'.

He calls the line of apses 'quiescent' for $n = 1$, and 'moving' for $n \neq 1$. As the body orbits going from the upper to the lower apsis, the apsidal angle can be more or less than π, depending on the nature of the central force.

When $n < 1$ there occurs positive precession in which the orbit advances. This is equivalent to having a central force falling off more rapidly than the inverse square law. The exponent of the force is less than -2, and this is what Hall found in the case of the precession of Mercury in 1895.

Alternatively for $n > 2$, the orbit retards or shows negative precession. This is indicative of a force falling off less rapidly than the inverse-square. It implies that the exponent in the force law is greater than -2. In this manner, an empirical determination of the angle of apsis can determine the nature of the exponent in the central force law.

In Book III of the *Principia*, Newton argues that the force between the sun and the planets is inverse-square precisely because the line of apses appeared to be 'quiescent.' However, during the course of the following century, in which the time lapse of planetary observations increased, it became evident that the orbits of the planets were slowly precessing—so much so that Laplace chose not to use Newton's theorem to infer the inverse-square law for the planets.

Newton's theorem of revolving orbits is more general than 'just' explaining apsidal precession. According to his theorem that addition of an inverse-cube central force can produce a rotating orbit in which the angular speed is increased by a factor m while the radial motion remains unchanged.

Yet, Newton's theorem is more general than accounting for apsidal precession. For it describes the effect of adding on an inverse-cube law to any central force: If $r = r(\phi)$ is an orbit of angular momentum L under the action of *any* central force $F(r)$, then $r = r(m\phi)$ is an orbit of angular momentum mL under the action of a central force:

$$F_2(r) = F_1 - (m^2 - 1)\frac{L_1}{r^3}, \tag{2.132}$$

where L_1 is the magnitude of the angular momentum, of body 1, per unit mass. For $m^2 > 1$, the added inverse-cube force is attractive, while for $m^2 < 1$ the added force is repulsive.

The proof of (2.132) is quite simple. The only assumption is that the angular speed of the second body is m times that of the first, viz.,

$$\dot{\phi}_1 = m\dot{\phi}_2.$$

In terms of their angular momenta, this means that $L_1 = mL_2$ since the two radii are the same. The equations of motion of the two bodies are also identical:

$$\ddot{r} = F_1(r) + \frac{L_1^2}{r^3} = F_2(r) + m^2\frac{L_2^2}{r^3},$$

where, as usual, both F_i and L_i are referred to a unit mass. Rearranging the above expression gives (2.132).

If the angle variables of the two bodies are related by $\phi_1 = m\phi_2$, and if the path of the first particle is an ellipse with semi-latus rectum p and eccentricity, ε,

$$r = \frac{p}{1 + \varepsilon \cos \phi_1},$$

then the orbit of the second body will be given by:

$$r = \frac{p}{1 + \varepsilon \cos(m\phi_2)}.$$

If the elliptical orbit is 'quiescent', in Newton's terminology, the body will rotate about the center of force by π without a change in the radial motion. If the orbit is in 'motion,' the body will rotate about the center of force as it moves from one apsis to the other. The corresponding apsidal angle will be π/m.

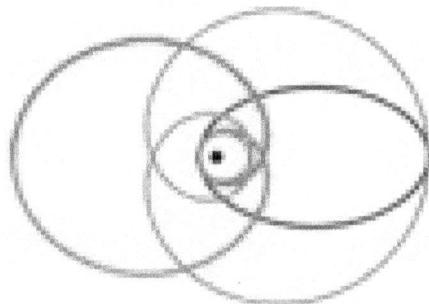

Figure 2.31: Higher order harmonic orbits: $m = 1$ in blue is the fundamental, $m = 2$ is in magenta and $m = 3$ is in green. Taken from https://ipfs.io/ipfs.

Newton also emphasized that the radial motion $r(t)$ for both orbits is the same; consequently, if the new orbit is viewed from axes rotating at a rate $(m - 1)\dot{\phi}$, then the two orbits would have the same form. However, the rotation rate $(m - 1)\dot{\phi}$ is a constant only for circular orbits. For elliptical orbits, the rotation rate increases near the perihelion and decreases near the aphelion. Whereas, if we view the orbit from axes that are uniformly rotating, the shape of the two orbits can differ enormously. [51]

If m is close to 1, the second orbit resembles the first, but revolves slowly about its center of force. The precession is in the same direction for $m > 1$, while it precesses in the opposite direction for $m < 1$.

Moreover, Newton's theorem shows that the inverse-cube law may be applied to linear and inverse-square laws, while without jeopardizing their closure, it causes them to precess. If $m = \sqrt{n}$ is an integer number, different from 1, it will complete m rotations in the time it takes the fundamental, $m = 1$, to complete a single rotation, as shown in Fig (2.31).

In contrast, for m a fraction, subharmonic orbits occur which although they show weirder behavior, as seen in Fig (2.32), they still do not violate Bertrand's theorem.

---

[51]D Lynden-Bell & R M Lynden-Bell, "On the shapes of Newton's revolving orbits," *Note Rec R Soc Lond* **51** (1997) 195-198.

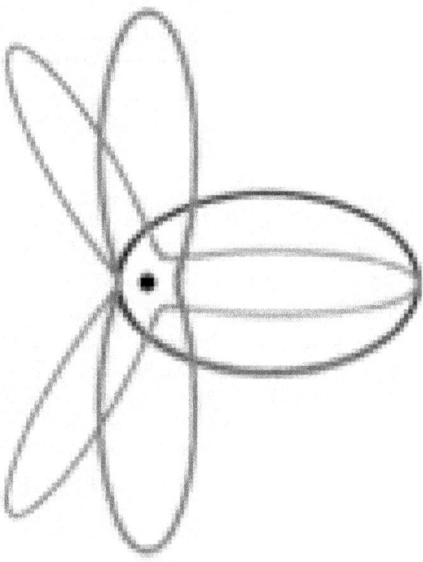

Figure 2.32: The harmonic orbit, $m = 1$ (blue), and some subharmonic orbits for $m = 1/3$ (magenta) and $m = 1/3$ (green). Taken from https://ipfs.io/ipfs

Amount (arc-sec/century)	Cause
5025.6	Coordinate (due to precession of equinoxes)
531.4	Gravitational tugs of the other planets
0.0254	Oblateness of the sun (quadrupole moment)
42.98±0.04	General relativity
5600.0	Total
5599.7	Observed

Figure 2.33: The putative sources causing the perihelion of Mercury. They are assumed to be independent–and additive. Unlike the other components, the 43 arcsec is attributed generically to "general relativity". This is wholly unsatisfactory as an explanation!

In Proposition 2 of Book III, Newton applies his theorem to Mercury, with eccentricity 0.2056, and even suggested that it may apply to Halley's comment with an eccentricity of almost 0.97. This was over a century prior to Le Verrier's observation that there was some 38 arcseconds per century in the precession of Mercury's apse that could not be accounted for by the inverse-square law. This estimate was revised by Simon Newcomb in 1882 bringing it to 43 arcsec per century. The individual contributions are shown in Fig (2.33).

Taking the force law to be of the form $-\mu r^N$, Asaph Hall in 1894 made the hypothesis that there was a deviation in the inverse-square law, and applied Newton's formula in the form $\pi/\sqrt{(N+3)}$, where $N+3$ was Newton's n. Thus, given Newcomb's value of the precession, Hall deduced an exponent of $N = -2.00000016$ for the force law exponent. This is a minute addition of $1.6 \cdot 10^{-7}$ to Newton's inverse-square law.

It is commonly believed that Hall's approach of using Bertrand's formula, for he was unaware of Newton's theorem of rotating orbits, was not totally convincing, and the so-called 'mystery' of Mercury's prograde precession had to wait for Einstein and his general theory of relativity. That this theory also determines the precession of the other planets is false. Venus, Uranus and Pluto all retrograde, and there is nothing in the Schwarzschild metric that would allow for the conversion of a force that decreases more rapidly than the inverse-square into one that decreases more slowly.

To derive the expression for precession, we turn to the work of Bertrand [52] which

---

[52] J Bertrand, "A theorem relative to the motion of a point pulled towards a fixed centre," *C R Acad Sci Paris* bf77 (1875) 849.

proves that only potentials which give rise to closed orbits are the Newtonian and Hookean potentials.

Introducing (2.132) into (2.61) gives:

$$u'' = -u + \frac{\mu}{L^2} - (m^2 - 1)u.$$

Canceling terms, it reduces to:

$$u'' = -m^2 u + \frac{\mu}{L^2}. \tag{2.133}$$

There is no need to consider nearly circular orbits since (2.133) is linear. Hence, any force law with m rational will give closed orbits.

If we introduce a dipole force, as a prototype of a tidal force ,

$$F = \frac{\mu}{r^2} - \frac{\mu d}{r^3},$$

where d is the distance between fixed bodies, or a distance traversed over which the magnitude of gravitational attraction changes, then we find:

$$m^2 = 1 - \frac{\mu d}{L^2}. \tag{2.134}$$

Since $m < 1$, the force falls off more rapidly than the inverse-square, a result already obtained by Hall.

To obtain an expression for d, let us now compare it to Einstein's modified Keplerian ellipse,

$$u'' = -u + \frac{\mu}{L^2} + 3\frac{\mu}{c^2}u^2, \tag{2.135}$$

that we discussed in §1.3.2. Considering nearly circular orbits, we expand the right-hand side of (2.135) about its roots:

$$u_\pm = \frac{1 \pm \sqrt{1 - 12(\mu/cL)^2}}{6(\mu/c^2)} \simeq \frac{1 \pm [1 - 6(\mu/cL)^2]}{6(\mu/c^2)}. \tag{2.136}$$

For the smaller root, u_- we find to lowest order:

$$x'' = -m^2 x + O\left(x^2\right), \tag{2.137}$$

where $x = u - u_-$, and

$$m^2 = 1 - 6 \left(\frac{\mu}{cL} \right)^2 . \tag{2.138}$$

The apsidal angle is, therefore,

$$\phi = \pi/m \simeq \pi \left\{ 1 + 3 \left(\frac{\mu}{cL} \right)^2 \right\} . \tag{2.139}$$

Now,

$$6 \frac{\mu}{pc^2} = 6 \frac{2.19 \cdot 10^{13}}{5.5 \cdot 10^{10} c^2} = 1.77 \cdot 10^{-7}$$

which is to within 6.6 % of Hall's result $\delta = 1.6 \cdot 10^{-7}$, obtained from Newton's formula for the apsidal angle.

However, the apsidal angle should be a sole function of the eccentricity, and certainly not terms extraneous to the ellipse. Moreover, it should increase with increasing eccentricity. Thus, it became a more sensitive index of deviations from the inverse-square law.

Hall's attempt to account for the 43 arcseconds was supposedly ruled out by Ernest Brown's development of the Hill-Brown lunar theory in 1903. The final deathblow supposedly claim with Einstein's 1915 derivation using his general theory.

Comparing (2.134) with (2.138) we obtain:

$$\frac{\mu}{d} \sim c^2, \tag{2.140}$$

showing that d is of the order of the Schwarzschild radius. By considering weak gravity in the solar system where $r >> \mu/c^2 \sim 1.5$ km, all relativistic effects are indeed negligible. Moreover, the total energy of bound planetary orbits per unit mass are much smaller than c^2. Furthermore, in comparison with planetary speeds, the speed of light is practically infinite. Then what type of equilibrium does (2.140) describe? Is there really a planet called Vulcan?

The justification of a result cannot be based merely on a numerical coincidence. Just recall that the Ptolemaic theory of cycles and epicycles were concocted to reproduce the planetary orbits–without any physical justification for doing so!

To say that the sun with its enormous mass "creates its dent in space-time [so that] Mercury, so firmly embraced by our star's gravitational field, lies deep within that solar gravity well" [53] begs the question.

For according to Einstein's theory, the 'dent' exists whether or not mass is present. Condition (2.140) requires an exceptionally strong gravitational field that would place it some 3000 meters from the sun, and not "at a distance of about twenty-one million miles from the sun, with a period of 34 days and 16 hours," as claimed by Stephen Alexander of Princeton who re-worked Le Verrier's calculation.

It is also surprising that there is no connection between Newton's theorem of revolving orbits and Einstein's prediction. It's like both exist on separate planets. The equilibrium condition (2.140) is enough to dissuade one to look for a physical explanation in the general relativistic calculation. It is not as Levenson claims that "Einstein's pen destroyed Vulcan and re-imagined the cosmos."

As a final comment, we remark that if we had chosen the larger root in (2.136), which is approximately,

$$u_+ \simeq \frac{1}{3}\frac{c^2}{\mu} - \frac{\mu}{c^2},$$

we would have obtained:

$$m^2 = -\left[1 - 6\left(\frac{\mu}{cL}\right)^2\right],$$

corresponding to a Cotes spiral, known as a Poinsot spiral. The solution to (2.137) is:

$$x = x_0 \cosh\left\{\left[1 - 6\left(\frac{\mu}{cL}\right)\right](\phi - \phi_0)\right\},$$

where x_0 and ϕ_0 are two constants of integration. The form of the Poinsot spirals are shown in Fig (2.34). They correspond to an unstable solution, where Mercury would plunge into the sun, or escape to infinity, and, consequently, is excluded from further consideration.

Poinsot spirals belong to a general class of Cotes spirals, defined in (2.62). Out of this class of spirals, Newton singled out the equiangular spiral, as being produced by an

---

[53]T Levenson, *The Hunt for Vulcan* (Random House, NY, 2015).

Figure 2.34: Poinsot spirals corresponding to $m = 1$ in green, $m = 1/3$ in cyan, and $m = 1/6$ in blue. Taken from https://ipfs.io/ipfs.

inverse-cubic force. The equiangular spiral has a constant angle between the radius vector and the tangent to the curve at any point along the trajectory. The motion is unstable, either spiralling into the center or out to infinity, unless the initial conditions are just right that the orbit closes. This will occur, in general, with the addition of an inverse-square law, as we have seen. Moreover, Newton was able to generalize this by proving that the force law determines the character of the curve, and conversely. This is something that general relativity negates.

2.16 Precession beyond inverse-square

With the discovery of flat rotation curves of spiral galaxies, and the recognition that bars and spirals are dominant components of spiral nebulae, as can be seen in Fig (2.35),

Disc instabilities is a major mechanism for forming stars, planets andBar.jpegGalactic discs form dense clouds which then condense to stars. Another type of disc instability

Figure 2.35: The bar is embedded at the center of the galaxy's spiral arms. The myth surrounding it is that it cuts across the area where a "supermassive" black hole resides. On the contrary, it is the "central engine" which sucks in matter from the dust torus causing it to radiate with a broad *non-thermal* spectrum. The bar consist of relatively old and red stars, measuring some 27 K light years long.

is *bar* formation. A bar through the center of a spiral galaxy thereby acquires potential energy. It is possible that the spiral arms of the galaxy become wrapped up through rotation. They may, as well, explain why stars come in binaries through the condensation on either side of the bar. [54]

Their discovery prompted attention being called to self-similar power laws and a force law that varied as the inverse-power of distance, i.e., the logarithmic potential. A generalization,

$$\varepsilon = \frac{r_a^k - r_p^k}{r_a^k + r_p^k},$$ (2.141)

of the eccentricity:

$$\varepsilon = \frac{r_a - r_p}{r_a + r_b},$$ (2.142)

where r_a and r_p are the radial distance from the center of the aphelion and perihelion, respectively, was used because it reduced to the usual definition of ε for an ellipse for $k = 1$, and it seemed to be a good approximation for the planets.

Newton, however, could not predict the gap between known perturbations and the observational value of the advance of the perihelion of Mercury. In fact, in Proposition 2 of Book III, he applies his formula (2.131) to the orbit of Mercury. The eccentricity of Mercury is 0.2056, and the greatest solar distance is 1.5 times that of its least distance so that expression (2.142) is a very good approximation to the true value of the eccentricity. It was the 43 arc-seconds that escaped Newtonian mechanics that propelled general relativity to almost universal acceptance. Numbers work magic!

In another application of his formula, Newton claims that the force between the Earth and the Moon is inverse-square. In Proposition 3 he argues that the lunar apsis moves, on the average, 3.3 arc-minute per revolution so that the exponent in his force law is $n - 3 = -2 - (4/283)$, which gives a force law exponent $n = 2.967$, falling off slower than inverse square. This he attributes to the radial perturbing force of the Sun.

This is at least a physical explanation, not like general relativity which attributes the gap in Mercury's precession simply to general relativity (c.f. Fig (2.33)). But, whatever it is, it is apparently caused by a force that falls off *faster* than the inverse square since

---

[54]W Kundt, *Astrophysics: A Primer* (Springer, Berlin, 2001), p. 79.

it predicts $n = 1 - 6(\mu/L)^2$. However, numerical equivalences do not explain physical causes. And even the numerical coincidence is in doubt since the precision to which the current emphemerides is still largely unknown. [55]

The logarithmic potential stradles both hyperbolic orbits giving deflection and elliptic orbits that precess. Recall that the radial orbital differential equation is given by Eq (2.61), where the generic function, $F(1/u)$, is $-dV/dr$ with V as the scalar potential. In the logarithmic case, the potential V is given by:

$$V(r) = \ln\left(\frac{K}{r}\right), \tag{2.143}$$

where K is a constant.

Introducing (2.143) into (2.61), and rearranging give:

$$uu'' + u^2 = \frac{K}{L^2}. \tag{2.144}$$

The orbits have the general form of ellipses, and have a long history dating back to Newton's *Principia*, as we have seen.

A rather good approximation to orbits governed by logarithmic and power potentials is: [56]

$$kz \equiv (pu)^k = 1 + \varepsilon \cos(m\phi). \tag{2.145}$$

For $k = 1$, (2.145) corresponds to the inverse-square law, while for $k = 2$, the orbit will be bound, but not closed. This is a hallmark of the logarithmic potential. The graph of (2.145) is shown in Fig (2.36), where the orbit is bound but not closed. For the logarithmic potential, the azimuthal frequency, $\dot{\phi} = L/r^2$, will be given by:

$$\dot{\phi} = \frac{L}{p^2}\left[1 + \varepsilon \cos(m\phi)\right],$$

<hr/>

[55] A A Vankov, "General relativity problem of Mercury's perihelion advance revisited," arXiv:1008.1811 [physics.gen-ph].

[56] C Struck, "Simple, accurate, approximate orbits in the logarithmic and a range of power-law galactic potentials," *AJ* **131** (2006) 1347-1360.

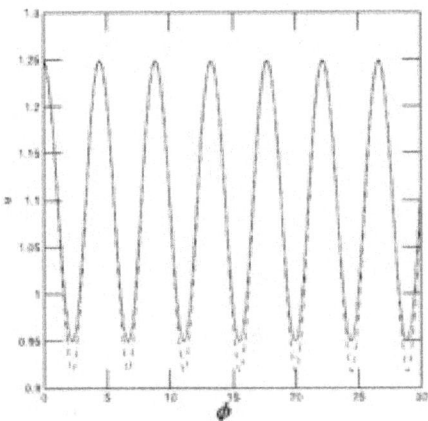

Figure 2.36: The inverse radius versus the azimuth. The solid line is a numerical integration of the orbit, while the dashed line is the first order p-ellipse approximation, valid to order e. Taken from Struck, *op. cit.*

which is one of the few cases where the orbital period T can be found in closed form. Integrating, we find:

$$\frac{L}{m\sqrt{1-\varepsilon^2}}2p^2t = \tan^{-1}\left(\sqrt{\frac{1+\varepsilon}{1-\varepsilon}}\tan(m\phi/2)\right),$$

where we have set the initial azimuth, ϕ_0, at the initial time $t = 0$, equal to zero. This corresponds to the periapse. Thus, setting $m\phi = \pi$, corresponding to the apoapse, so that the orbital period, T, is found to be:

$$T = \frac{\pi p^2}{Lm\sqrt{1-\varepsilon^2}}. \tag{2.146}$$

The period depends on the eccentricity through the factor $m\sqrt{1-\varepsilon^2}$. This factor will make its way into the expressions of an associated ellipse via the semi-latus rectum, p.

Rearranging (2.144) we get:

$$uu'' + u^2 = \frac{K}{L^2}.$$

Introducing (2.145) converts it into:

$$2z - \frac{m^2(1-\varepsilon^2)}{2z} = m^2 + p^2 K.$$

For

$$m = \frac{\alpha}{\sqrt{1-\varepsilon^2}} \tag{2.147}$$

with $\alpha = \text{const.}$, we recognize:

$$w = \frac{1}{2}\left(z + \frac{\alpha^2}{z}\right) \tag{2.148}$$

as a generalization of Joukowski transform, (2.55), for $\alpha = 1$.

It is interesting to observe that (2.147) corresponds to the second-order expression found in Ref. [57], for the particular value $\alpha = \sqrt{2}$ when $\varepsilon \cos(m\phi)$ is expanded in powers of the eccentricity, and the approximation, $\sqrt{1-\varepsilon^2} \simeq 1 - \varepsilon^2/2$ is used.

As we have already remarked, values of $n > 1$ are indicative of force laws that fall off less rapidly than the inverse-square law, and so the orbit will retard, or retrograde. The azimuth must increase less than 2π in order for the system to return to its original position.

With $z = x + iy$ and $w = u + iv$, the mapping from the z to the w plane consists in transforming a circle of radius $r > a$ into a Hookean ellipse whose center coincides with the origin, and whose semi-major and semi-minor axes are given by $a = r + \alpha^2/r$ and $b = r - \alpha^2/r$, respectively.

We now use the property illustrated in Fig (2.22) that when the Joukowski ellipse is squared we get another Joukowski ellipse shifted a distance of $2\alpha^2$:

$$w^2 = (2z)^2 + \left(\frac{\alpha^2}{2z}\right) + 2\alpha^2.$$

The first two terms correspond to an origin center ellipse with foci at $\pm 2\alpha^2$; the last term shifts the left-hand focus to the origin. In this way a Hookean ellipse has been transformed into a Newtonian ellipse.

---

[57] Struck, *op. cit.*

It seems like by a mere flick of a magic wand by which the focus is translated to the
origin without affecting the orbit. But what has changed is the nature of the attracting
force. When the attracting point was at the center we had a Hookean ellipse with a
linear restoring force, yet by a 'mere' translation of the focus to the origin we get an
inverse-square law—with the same identical orbit! This is just a confirmation of what
we already discovered in §1.3.1, but there is more.

The eccentricity of the ellipse,

$$\varepsilon = \frac{2m\sqrt{1 - \varepsilon^2}}{r + m^2(1 - \varepsilon^2)/r},$$

can be recast in the form:

$$\frac{1 - \varepsilon}{1 + \varepsilon} = \left(\frac{r - \alpha}{r + \alpha}\right)^2, \tag{2.149}$$

which bears a striking similarity to (2.58).

Now, the left-hand side of (2.149) is just the ratio of the extremes of the orbit:

$$\frac{r_{per}}{r_{ap}} = \left(\frac{1 - \varepsilon}{1 + \varepsilon}\right), \tag{2.150}$$

where r_{ap} is the aphelion, and r_{per} is the perihelion. If we interpret ε as a relative speed,
then (2.150) is just a product of longitudinal Doppler shifts that we would normally get
upon reflection. [58]

We may develop this analogy further. The eccentricity is defined as:

$$\varepsilon = \frac{r_{ap} - r_{per}}{r_{ap} + r_{per}}. \tag{2.151}$$

If we emit a light signal from one of the apses and is reflected back at the other apsis,
the distance covered is $(1/2)(r_{ap} - r_{per})$. While, according to Einstein's method of
synchronization, time of propagation of the signal is the arithmetic mean of the forward and backward journeys, $(1/2c)(r_{ap} + r_{per})$, divided by the speed of light. Their
ratio is simply proportional to (2.151). Thus, we have *motion without motion*.

---

[58] B H Lavenda, *A New Perspective on Relativity: An odyssey in non-Euclidean geometries* (WS, Singapore, 2011) Ch 8.

2.17 Time delay in three and two dimensions

With a lapse of over half a century, Irwin Shapiro took Einstein at his word, and seriously considered the fact that "the speed of light depends on the strength of the gravitational potential along its path." Shapiro measured the delay of radar signals being bounced off Venus. The maximum of 200 microseconds occurred when the sun passed between the line of sight from earth to Venus.

If x is in the direction of the line of sight, the travel time of a light signal is increased by an amount:

$$t + \Delta t = \int \frac{dx}{c_0(1 - GM/rc^2)} \approx \int \frac{dx}{c_0} + \frac{GM}{c_0^3} \int \frac{dx}{r},$$

where Einstein's expression has been used to express c in terms of its value in vacuum, c_0. When the sun is aligned between earth and Venus, the distance from the center of the sun to Venus is $r = \sqrt{R^2 + x^2} \approx x$, at large distances. The additional delay is therefore:

$$\Delta t = \frac{GM}{c_0^3} \ln r + \text{const.}$$

The increase in time is thus due to the presence of a $2D$ potential, whose coefficient is proportional to the linear mass density. We will shortly see that the deflection angle, or deficit angle, as it is commonly referred to, is determined by the gradient of the deficit potential, (2.128), which is none other than the gravitational potential in $2D$.

In $3D$, the increase in time to traverse a distance L and back in the presence of a Newtonian gravitational field is:

$$\Delta t_{3D} = \frac{2L}{c} \cdot \frac{GM}{rc^2}, \tag{2.152}$$

whereas, in $2D$, it is given by the Shapiro potential:

$$\Delta t_{2D} = \frac{2L}{c} \cdot \frac{G\xi}{c^2} \ln(r/r_0), \tag{2.153}$$

where ξ is the linear mass density, which will reappear in the next section.

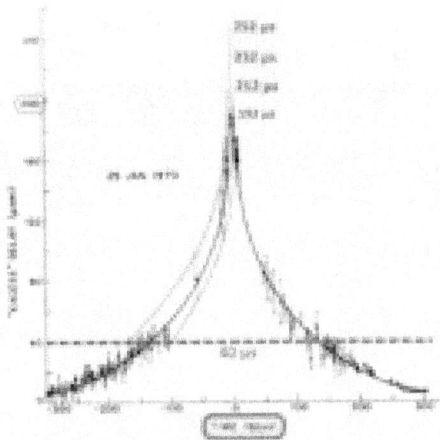

Figure 2.37: Shapiro's time delay measurements for a radar signal being bounced off Venus.

Whereas the Newtonian time delay, (2.152), is a *local* time delay, decreasing with r as we go further away from the deflecting body, the Shapiro time delay, (2.153), actually increases with distance!

Shapiro's results are displayed in Fig (2.37). The maximum delay of 200 μ-seconds occurs when the sun is in the direct path between Venus and earth, corresponding to the time 0 in the diagram. As the earth and Venus move on, the radar signal bypasses the sun, the time delay decreases. It will have decreased to around 40 μ-seconds when the sun is 100 times further away, which is 20% of the maximum value. This is to be contrasted with the 1% of its value for the bending of light at the same distance.

Thus, a 'mere' change of dimension converts a *local* effect into a *long-range* one. Yet, most, if not all, texts on special relativity hail Shapiro's experiment as a validation of Einstein's formula to within 3%!

By Gauss's theorem states that the product of the field over the boundary of a 'ball' must be constant. In $3D$, the field decays as $1/r^2$, and the 'ball' is a sphere, and the measure of its boundary is its surface area, $4\pi r^2$. The product of the two is a constant. In $2D$, the 'ball' is a circle, and a measure of its boundary is its circumference, $2\pi r$.

Figure 2.38: The monopole (left), dipole (center), and quadrupole (right) terms in the $2D$ multipole expansion. Taken from Tsogtgerel Gantumur, "Multipole expansion in the plane".

The field falls off as $1/r$, and their product is again constant.

Both potentials tend to infinity as $r \to 0$. But, whereas Newton's potential tends to zero in the limit $r \to \infty$, the Shapiro potential becomes unbounded in the same limit. Hence, there is the necessity to install an upper-cutoff, like the boundary of a disc.

2.18 Lower dimensional gravity

If we ask what phenomena depend on a linear inverse law, we see that they must occur in $2D$. In $2D$ there is no axis of symmetry, and the multipole expansion is:

$$\Phi = a_0 \ln \frac{1}{r} + \frac{a_1 \cos \phi + b_1 \sin \phi}{r} + \frac{a_2 \cos 2\phi + b_2 \sin 2\phi}{r^2} + \cdots , \qquad (2.154)$$

where a_i and b_i are coefficients. The first term is the monopole, which in $3D$ has a $1/r$ dependency, the second term is a dipole term, which in $3D$ would have a $1/r^3$ dependency, and the third term is a quadrupole, which would have a $1/r^5$ dependency in $3D$. They are shown in Fig (2.38). Consider a gravitational lens which is produced by a cluster of spiral galaxies's tremendous gravitational field. Such a gravitational field can bend light that will brighten, distort, and multiply the image of a more distant cluster.

The bending of light rays can occur when light passes from one medium to another with a different index of refraction as well as when it passes from the vacuum into a gravitational field. Possibly the simplest object that shows intrinsic curvature is a cone, which has a curvature singularity at its apex. If we consider the apex as the analogue

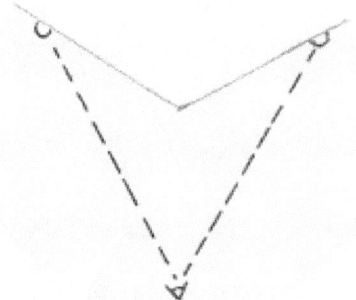

Figure 2.39: Sitting a cone through its apex, flattening without crinkling or creasing, and removing the wedge produces two light beams to the same object for an observer situated near the mass singularity.

of mass, we have a very simple model of geometrical lensing. [59]

A cone has a 'conical' defect which can influence geodesic motion. Imagine a $2D$ creature inhabiting a $2D$ universe which consists of the surface of the paper cup. If the singularity is very far away, he or she believes that the world is flat. The same is true of light rays for they will travel in straight lines without being bent.

Now cut a slit in the cone along a line passing through the apex, and flatten it out without creasing or crinkling. What is left is a flat disc with a wedge missing. We denote the 'deficit angle' by α. . The inhabitants would not notice any difference, and would still believe their world is flat.

Light rays too would not notice anything different than travelling in a flat plane, far from the cone point of course. Now place our observer near the mass singularity where his line of sight passes close to the singularity. However, because of the slit, the observer will now see two images of the same object, one on each side of the singularity, as shown in Fig (2.39).

---

[59]T Phillips, "Gravitational lensing and geometic lensing," *AMS* 1999.

The double image formation is a form of geometric lensing.

The density of spiral galaxies falls off as a power law:

$$\rho(r) = \rho_0 \left(\frac{r}{r_0} \right)^{-\beta}, \qquad (2.155)$$

with $\beta < 3$, which is related to the effective dimensionality. Introducing (2.155) into Kepler's III, we get:

$$v(r) = \left(\frac{4\pi G \rho_0 r_o^\beta}{3 - \beta} \right)^{1/2} r^{1-\beta/2}.$$

Now, using the fact that for many spiral galaxies, the rotational speeds are constant, we fix $\beta = 2$ and get:

$$\xi \equiv= \rho r^2 = \rho_0 r_0^2, \qquad (2.156)$$

which is a constant, linear mass density. This brings to mind the putative 'cosmic strings,' which would be long filaments normal to the galactic plane. [60] The deficit angle is related to the density of matter by:

$$\alpha = 4\pi G \xi.$$

The replacement of the point mass by its density can be considered as a 'regularization' of the matter source.

If Newton's law holds on all scales, it become necessary to invoke invisible, or 'dark,' matter in quantities that increase radially with r in galaxies. The ratio of mass to light, M/L, is used to measure the difference between, M, mass and, L, luminosity in galaxies. For ordinary stars, the ratio is unity, but, in spiral galaxies it is much larger, thereby requiring the presence of 'dark' matter.

Another way to modify Newton's law of gravitation is to assume that the mass increases with r over large scales. In Fig (2.40), which is taken from Wright, Disney and Thomson, [61] the ratio M/L is plotted versus r, the linear scale-size. There is a general trend towards linearity over three orders of magnitude in scale.

---

[60]A Vilenkin & E P S Shellard, *Cosmological Strings and other Topological Defects* (Cambridge U P, Cambridge, 1994).
[61]Wright, *idem.*

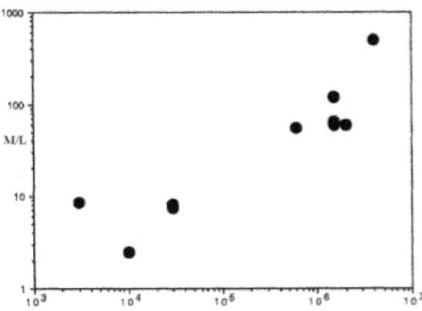

Figure 2.40: Mass-to-light, M/L, ratio plotted versus the linear scale-size, r, in light-years.

If, in fact, M/L is a linearly increasing function of r, then the asymptotic velocity, (2.124), would not be constant, and MOND would not reflect the flat velocity rotational curves at small acceleration.

We can distinguish between $M(r)$ and L by the equations they obey. Assuming the former to satisfy Newton's inverse square law:

$$\frac{v^2}{r} = \frac{GM}{r^2},$$

while the latter to be given by the inverse linear law:

$$\frac{v^2}{r} = \frac{GL}{r r_0},$$

at least for $r \gg r_0$, then equating the two results in: [62]

$$\frac{M}{L} = \frac{r}{r_0}. \tag{2.157}$$

The mass-luminosity relation, (2.157) is analogous to the density relation, (2.156).

A cone is flat precisely because it is created by rolling up a flat sheet. The act of rolling up preserves the metric on the interior of the sheet, not on the boundaries where the

---

[62]Wright, *idem.*

wedge was cut. You only get curvature when you crinkle or crumple a flat sheet. When a wedge α is cut out of a piece of paper the line element is:

$$ds^2 = dr^2 + r^2 d\phi^2,$$

where the azimuthal angle ϕ is now constrained to the closed interval, $0 \leq \phi \leq 2\pi - \alpha$.

Alternatively, we can choose a scaled angle, $k\phi = \varphi$, where $k = 1 - \alpha/2\pi$, and write the line element as:

$$ds^2 = dr^2 + \left(1 - \frac{2G\xi}{c^2}\right) r^2 d\varphi^2,$$

where the angular coordinate, φ, is again periodic with period 2π, and the angle has been normalized to 2π. The total angle of the conical singularity is $2(\pi - G\xi/c^2)$, which is the difference of a circle of radius 1 and a small circle centered about $r = 0$.

Light rays coming from the same original source may be entering our telescope from two different directions that differ by a deficit angle, α. This angle is related to Gaussian curvature, κ, in the following way. According to the Gauss-Bonnet theorem:

$$\frac{1}{2\pi} \int_S \kappa dS = \chi - \frac{1}{2\pi} \int_{\partial S} k_g d\ell, \tag{2.158}$$

over a surface S whose boundary is ∂S, where χ is the Euler characteristic, and k_g is the geodesic curvature.

Now, make a loop around the apex. The Euler characteristic, $\chi = 1$, showing that region is homeomorphic to a disc. The geodesic curvature of the loop outside the cone is the same as in Euclidean space, $k_g = 1/r$, and $d\ell = rkd\alpha$. Introducing these values into the Gauss-Bonnet theorem in $2D$ results in:

$$\int_S \kappa dS = \alpha, \tag{2.159}$$

which shows a non-vanishing curvature inside ∂S. A vector under parallel transport about the origin will be rotated by an angle α.

A negative angle would mean that the total angle of the conical singularity is greater than 2π. Instead of positive curvature, for values of $\alpha < 2\pi$, we now have negative curvature. At the apex of the cone there is a negative mass density implying that gravity is now repulsive. This requires a negative linear mass density, and shows up when two parallel geodesics passing the particle at the center of the disc are bent away from each other, or, in other words, they are repelled from the singularity.

The gravitational potential causing the lensing is the $2D$ potential:

$$\Phi_S(r) = G\xi \ln(r/r_0), \tag{2.160}$$

which we will refer to as the Shapiro potential. It is the first term in the $2D$ multipole expansion (2.154). The corresponding gravitational force is:

$$g(r) = -\frac{G\xi}{r}, \tag{2.161}$$

which is comparable to the second term in (2.125). In that expression, M/r_0 plays the role of a constant, linear mass density, ξ.

The modification of the Newtonian gravitational force, (2.161), is not the only case where there is a modification of the gravitational force. Zwicky [63] reasoned that a photon passing near a massive body will not only be deflected, "but it will also transfer momentum and energy to the mass, and make it recoil. During this process, the light quantum will change its energy and, therefore, its frequency."

Zwicky thought that this would be the gravitational analog of the Compton effect:

$$z = \frac{2\hbar\omega_0 \sin^2(\Theta/2)}{m_e c^2}, \tag{2.162}$$

where Θ is the scattering angle, and m_e the mass of electron. The shift due to Compton scattering, (2.162), would make the z-number dependent on the wavelength of the incident relation through ω_0. This renders it dubious as to whether it could explain the large red-shifts observed in the spectra of extra-galactic objects.

Zwicky also introduced the concept of gravitational 'drag,' or as it is now commonly referred to as 'tired' light. The Newtonian potential would be modified as:

$$\Phi_Z = -G\rho Dr, \tag{2.163}$$

---

[63]F Zwicky, "On the red shift of spectral lines through interstellar space," *PNAS* **15** (1929) 773-779.

where ρ is the uniform, average, density of the universe, and D is the gravitational 'drag,' which "should be as large as the dimension of the space over which masses are distributed, if those masses are regarded as independent of one another." But, the masses are coupled by their gravitational forces that they exert on one another so distance D must be very large with respect to clumped galactic matter.

Light of frequency, ν, traveling distance, r, would lose energy causing a red-shift of:

$$\frac{\Delta \nu}{\nu} = \frac{1.4 G \rho D r}{c^2} \sim 10^{-3}.$$

$D \gg r$ measures the distance over which the gravitational 'drag' operates. This provides an estimate of the Hubble parameter since:

$$\frac{\Delta \nu}{\nu} = \frac{H}{c} r.$$

The case of linear, density, (2.156), would correspond to the term $\rho D r$, and the red-shift would only increase logarithmically, and not linearly with r. This would break the chain in going from red-shift and velocity, via Doppler, and Hubble's velocity-distance relation.

Just as our paper cone has a circle as its base, a $3D$ cone has a sphere as its base, like an ice-cream scoop. We will consider only right cones where points on a horizontal slice through the cone are all at the same distance from the apex. The position of an observer situated at the North pole of the sphere will have his or her position determined by the intersection of great circles. The key point is that the space which the observer is looking into is made up of $2D$ cones based on those great circles. Light rays that connect any two points on a cone must remain in that cone so that the observer will see a cumulative effect by sweeping over all $2D$ cones. Thus, a $1D$ field of vision has now become a $2D$ one. And since for each great circle there will be two images, equidistant from the cone's singularity, the combining of all the cones will turn these individual images into a halo that is symmetrical with the observer's line of sight when it passes through the cone's apex. Any motion on his or her part will cause a distortion of the halos, and eventually will make them disappear.

2.19 General Relativity in Lower Dimensions

It is well-known that GR fails to provide the proper Newtonian limit in $2 + 1$ dimensions. [64] This stems from the fact that in $2D$, the Riemann curvature is completely determined by the Einstein tensor. It is even worse in $1D$, where the Einstein tensor vanishes identically. According to Jackiw, [65] "gravity has to be invented anew since GR cannot even be formulated."

In $2D$, space-time must be flat where matter is absent. Consequently, there can be nothing like a Newtonian potential so planetary motion is out of the question, and, supposedly, GWs too. Since space-time is flat where mass does not tread, the geodesics of a test particle will be straight lines.

Since BHs are described by their areas, $2D$ should be an excellent place to characterize them. Taking our cue from Schwarzschild, the line element of a circularly symmetric space-time in is most general form is:

$$ds^2 = -e^{2B(r)}dt^2 + e^{2A(r)}dr^2, \tag{2.164}$$

where A and B are functions of the radial coordinate only. If they are independent, then a single Einstein equation is insufficient to determine them.

The most expedient procedure is to reduce the number of degrees-of-freedom to one by setting $B = A$. Thus, the line element reduces to a conformal one:

$$ds^2 = e^{2B(r)} \left(-dt^2 + dr^2 \right). \tag{2.165}$$

The scalar curvature, which is twice the negative of the Gaussian curvature in $2D$, is:

$$R = -2e^{2B(r)}B''(r),$$

where, as usual, the prime denotes differentiation with respect to the radial coordinate.

The field equation is:

$$B'' + \frac{B'}{r} = 0, \tag{2.166}$$

---

[64] See, for example, C Romero & F Dahia, "Theories of gravity in $2 + 1$ dimensions," preprint 23-93 DF/CCEN-UFPB.
[65] R Jackiw, *Nucl Phys B* **252** (1985) 343.

whose solution is simply:

$$B = 1 + a\ln(r/r_0),$$

where a and r_0 are arbitrary constants of integration. Now, for $B = 0$, the conformal metric reduces to the metric of Minkowski space-time. Thus, for small values of B, we can write $e^{2B} \simeq= 1 + 2B$, and introducing the value of B, enables the metric to be written as:

$$ds^2 = -2\left(1 + a\ln\frac{r}{r_0}\right)\left(dt^2 - dr^2\right).$$

Much effort has been expended in studying how BHs can form in gravitational theories of lower dimensions. The metric appears to possess a singularity at $r = 0$, but $r = r_0 e^{-1/a}$ does not appear to be an event horizon because light, $ds = 0$, is unaffected by it.

Writing down the geodesic equations and introducing them into the line element gives the conservation of energy equation for radial motion:

$$m\dot{r}^2 + G_2 M \ln(r/r_0) = \text{const.},$$

where we have set $a = G_2 M$, m the mass of the test particle, M the attracting mass, and G_2 the gravitational constant in $2D$.

If we go back to our original metric, (2.164), and do not use the fact that A is small, it too will solve the Einstein equation, (2.166), with $2A = -\ln(r_0/r)$. Introducing this into into the line element (2.164) now gives:

$$ds^2 = -2\left(1 + a\ln(r/r_0)\right)dt^2 + \left(\frac{r_0}{r}\right)dr^2. \tag{2.167}$$

And lo and behold we have a BH, referred to as a Kawai BH. [66] The singularity at $r = 0$ is now surrounded by an event horizon at $r = r_0 e^{-1/a}$. Although the Einstein field equation has been used to derive the coefficients in the line element, in $2D$, the full Riemann curvature can always be expressed in terms of the Ricci scalar, and if it

---

[66]T Kawai, *Phys Rev D* **48** (1993) 5668.

vanishes there can be no conical singularity at $r = 0$. Here is "where things start to unravel." [67] Although there is a BH, there is no Newtonian limit far from it.

In the original Kawai BH metric, the coefficients in (2.167) are squared. Then the coordinate substitution $x = r_0[1 + a \ln(r/r_0)]$ reduces it to what is known as a Rindler wedge:

$$ds^2 = -x^2 dt^2 + dx^2.$$

A Rindler wedge has zero Gaussian curvature, and a further coordinate transformation reduces it to the Minkowski metric. This led Cornish to ask where did the BH go?

To add fuel to the fire, all $2D$ spaces in GR are conformally flat so the most general metric is (2.165). With $B < 0$, *imaginary* time has a period $2\pi/B$, and an associated Hawking temperature, $T_H = -\hbar B/2\pi$. What makes it so special is that in $2D$, this temperature is proportional to its source, as opposed to being inversely proportional to it in $3D$. And finally, $-B$ has been associated as a 'proper acceleration' by Unruh. [68]

From all this it is more than sufficient to conclude that there is absolutely no physics in any of it.

2.20 Discordant red-shifts

Quasi-stellar objects, or quasars, have posed a problem for the big bang theory since their discovery over thirty years ago. Unlike normal stars, they emit a broad, non-thermal spectrum of radiation, and their emission lines are highly red-shifted; z-numbers of up to 4 and 5 have been reported in some cases.

Because the Doppler shift involves velocities, the high red-shifts of quasars have made them out to be very distant objects–that is according to conventional wisdom. Granted this, quasars are thought to contain valuable information about the structure of the early universe.

---

[67] N J Cornish, "The black hole that went away," arXiv:gr-qc/9609016.
[68] W Unruh, *Phys Rev D* **14** (1976) 870.

Luminosity has succumbed to red-shift by the establishment, contrary to observation, and this has led to the ruin of professional careers. It has also led to boycotting the publication of monographs. [69]

Quasars come under the banner of Active Galactic Nuclei (AGNs), which have very peculiar properties. Unlike normal stars which manifest spherical symmetry, AGNs have cylindrical symmetry with very dense object at their cores, much like the central massive objects in bars shown in Fig (2.35). The core has incorrectly been considered to a black hole, which it definitely is not.

It is comforting to know that Wolfgang Kundt [70] shares the same opinion:

> The galaxy feeds on an *active*, nuclear burning nucleus, a *burning disk*. . . Instead, most of my colleagues prefer to think of a supermassive black hole as the central engine of all the *active galactic nuclei*. They have not convinced me, after more than 25 years. AGN activity requires a refilling engine, with nuclear burning, magnetic reconnections, and explosive ejection of ashes. . . Black hole formation would require distinctly higher mass concentrations than are ever reached in galactic nuclei. . .

Concerning his reasons for his disbelief in black holes, he continues:

> Over 45 *black hole candidates* have been proposed during the past 30 years from the class of binary X-rays sources, both high-mass and low-mass — among them Cyg X-1 and A0620-00 — on account of their large mass functions, absence of strict periodicities, and absence of type-I bursts (understood as nuclear detonations at the neutron-star surfaces). To me, all of them look like neutron stars surrounded by massive ($\approx$ 5M$_\odot$) accretion discs, because of their often hard spectra up into the γ-ray range), highly structured, fluctuating light curves saturating at $L_{Edd}(1M_\odot)$ [Eddington luminosity] during outbursts, line-luminous wind zones, occasional jet-formation and/or super-Eddington behaviour, and because of

---

[69]H C Arp, *Quasars, Redshifts and Controversies*, (Interstellar Media, Berkely, 1987); *Seeing Red: Redshifts Cosmology and Academic Science* (Apeiron, Montreal, 1998)

[70]W Kundt, *Astrophysics: A New Approach*, 2nd ed. (Springer, Berlin, 2004), pp. 85-86, 111.

their indistinguishable further properties, as a class, from all the established neutron-star binaries. They just fill the gap between the high-mass and low-mass compact binaries.

Kundt concludes that for these reasons and more he shares "the doubts of a few other people, among them (the late) Viktor Ambartsumyan and [the late Fred] Hoyle, in the widely accepted black hole paradigm."

However, Kuhn misses the mark when he says that "the free-fall speed of an infalling test particle reaches the speed of light when crossing the hole's horizon. Consequently, a black hole's binding energy is of the order of its rest energy." This is definitely inaccurate: Because the motion is *geodesic* for which the velocity can change its direction, but not its magnitude, a radial test particle, together with light rays, arrives at the 'event' horizon with *zero* velocity, and *zero* coordinate and absolute accelerations. [71] Hence, a black hole's binding is *not* of the order of its rest energy, and the whole analogy with a Gibbs equation for black holes is nonsense. The putative expression for the entropy cannot be proportional to the square of the mass for that would violate the second law.

The radiation due to its gravitational attraction from the dust torus is definitely *non-thermal* [72] and so would violate Hawking's, again incorrect, assumption that black holes radiate with a thermal, black body, spectrum. In fact, Kundt [73] refers to the so-called Hawking temperature as "a somewhat speculative marriage of GR with field quantization." That marriage should have seen a divorce a long time ago, since Hawking's expression implies a negative heat capacity for a single phase system, and, consequently, it constitutes a violation of the second law of thermodynamics. [74]

The radiation of the core in turn will excite gaseous clouds that will emit line radiation, both narrow and broad. The model of non-thermal radiation by a central core, referred to as the central engine,[75] is shown in Fig.(2.41). The central core deflects light, and acts as a gravitational lens, just like the apex of a cone.

---

[71] Normally, we would require only the component of the acceleration tangent to the surface to vanish for geodesic motion. However, all three terms vanish under the condition that the radius reaches the Schwarzschild radius.

[72] D F V James, "The Wolf effect and the red-shift of quasars."

[73] Kundt, *op. cit.*

[74] B H Lavenda, *Thermodynamics of Extremes* (Albion, Chichester, 1995),

[75] D F V James, "The Wolf effect and the red-shift of quasars", arXiv: astroph: 9807205.

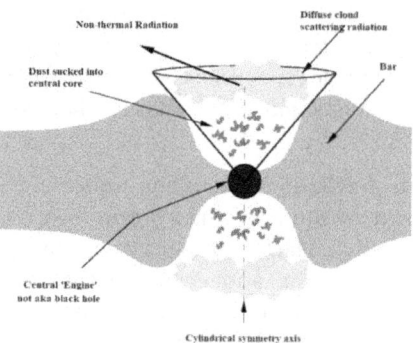

Figure 2.41: A model for the non-thermal radiation emission of a massive central core in an AGN. Adapted from James *op. cit.*

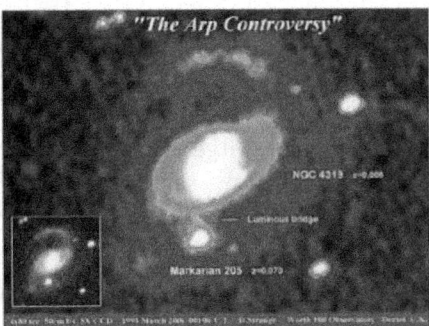

Figure 2.42: Known as the 'Arp Controversy,' the reality of the connection between galaxy NGC 4319 and the quasar/AGN Markian 205 was finally laid to rest by these CCD frames taken by D. Strange with a 50 cm telescope.

By employing the Wolf effect, the properties of AGNs can be explained by the observation of the same object in different directions. According to the Wolf effect, [76] the spectrum of light generated by two identical point sources depends not only upon their individual spectra, which are identical, but, in addition, upon the *spatial* correlations of the sources of radiation. This is a typical optical phenomenon where the cross term of the individual sources leads to a reddening of the frequency of the point sources.

What if the two identical point sources were created by gravitational lensing of close pair of the galaxy NGC-4319, with red-shift $z = 0.006$, and the quasar Mk-205, with red-shift $z = 0.070$, shown in Fig (2.42)? [77]

Despite having a huge difference in their red-shifts, their proximity says otherwise to their assumed huge difference in distances based on the assumption that the Doppler effect is the culprit. Instead of identical point sources, there would be identical images in different directions caused by the gravitational lensing of the quasar. Instead of random scatters, the lensing itself would cause a downshift in frequency from the frequency of the individual images.

---

[76] E Wolf, "Non-cosmological red-shifts of spectral lines," *Nature* **326** (1987) 363.

[77] Arp, *op. cit.*

Figure 2.43: On the left, in $2D$, is the potential generated by a mass at the origin is unbounded both as $r \to 0$ and as $r \to \infty$. On the right, in $3D$ and higher dimensions, the potential is unbounded only in the limit $r \to 0$. Taken from Gantumur, *op. cit.*

The effects of $2D$ gravitational lensing would be pronounced indeed: On large scales, gravitation would tend to an inverse-linear law instead of an inverse-square law. And since the Shapiro potential increases without bound in both limits $r \to 0$ and $r \to \infty$, there would be no 'open' universes, as illustrated in Fig (2.43). Computer studies have shown that tails and bridges can better be explained by a linear rather than a quadratic inverse law. [78]

Since the propagation time from source to observer varies from one image to another, there should definitely be a red-shift in the frequencies of their spectra. If the separation between multiple images is a few arc-minutes, the time delay between various light paths is in the range of hundreds to thousands of years. [79] We should, therefore, expect discordant red-shifts.

The time delays in $2D$ are determined by the Shapiro potential, (2.160). This is caused by the slowing down of light by a gravitational potential. The bending of light is considered to occur in $3D$; the two dimensions are mixed-up by attempting to apply Fermat's principle in the deriving the so-called deficit equation. [80]

The incompatibility of attempting to use the $3D$ Newtonian gravitational potential to derive the deficit equation requires it to be projected onto the plane containing the source, lens, and observer. This masquerades the fact that the time delay is entirely a $2D$ effect, and not a $3D$ one. The (3 space + 1 time) metric, like that of Schwarzschild's, is not applicable, and that explains why an integration is needed to get the Shapiro

---

[78]Disney, *opt. cit.*

[79]B Paczynski, "Will cosmic strings be discovered using the space telescope?" *Nature* **319** (1985) 567-568.

[80]c.f. K Kuijken, "The basics of lensing," in *Gravitational Lensing: A unique tool for cosmology* ASP Conference Series, Vol 30 (2003).

potential. Linear mass densities are involved, not surface densities, and this has made it fair game for 'cosmic strings,' as a topological defect responsible for gravitational lensing.

Suppose that the Shapiro potential, (2.160), transforms the 2-sphere metric into a circularly conformal metric:

$$ds^2 = \left(1 - 8\pi G\xi c^2 \ln(r/r_0)\right)\left(dr^2 + r^2 d\varphi^2\right). \tag{2.168}$$

Conformality transforms the ordinary dot product into a new geometric structure in the plane whose Gaussian curvature is:

$$\kappa = \frac{1}{2}\left(\frac{8G\xi}{c^2}\right)^2 \left[1 - \frac{G\xi}{c^2}\ln(r/r_0)\right]^{-3} > 0.$$

Its relation to a cone, can be gleaned through the transformation:

$$\left(1 - 8\frac{G\xi}{c^2}\ln(r/r_0)\right)r^2 = \left(1 - \frac{2G\xi}{c^2}\right)^2 r'^{\,2}.$$

For upon introducing this into (2.168), and to first-order in $G\xi$, we get:

$$ds^2 = dr'^{\,2} + \left(1 - \frac{\alpha}{\pi}\right)^2 r'^{\,2} d\varphi^2. \tag{2.169}$$

Consequently, *the cause of the time delay, characterized by the Shapiro potential, is caused by a topological defect of a mass at the apex of the cone which creates the azimuthal 'deficit angle'.*

3 General Relativity in the Realm of Geometrical Optics

How is it possible that that Einstein's modification of a Keplerian orbit (2.78) can give *both* the perihelion advance of a planet, and the deflection of light? One is a heavenly body whereas the other is massless. Neither the *particle*, nor the *photon* properties appear in the geodesic equation. Moreover, in their discussion of the motion in a centrally symmetric gravitational field, Landau & Lifshitz [1] abandon altogether the GR and opt for the simpler, and more physically transparent approach of geometrical optics (GO) via the Hamilton-Jacobi equation. . In this chapter we shall investigate the relationship between GR and GO.

GR generalizes the Minkowski line element:

$$-d\tau^2 = -dt^2 + dr^2 + r^2 d\sigma^2,$$

where $d\sigma^2$ is the element of a two-sphere to:

$$-d\tau^2 = -Bdt^2 + Adr^2 + r^2 d\sigma^2.$$

In a static, and isotropic, universe, the coefficients, A, and B, will be functions only of the radial coordinate, r. However, on account to the geodesic equations, they will also necessarily be a function of t because r is a function of t.

Scalar potentials would be expected to appear in A, and not B. But, because there are no forces in GR, they are relegated to the time axis, and so, to B. Moreover, disregarding the space part completely and interpreting the 'local', τ, and 'coordinate,' t times, as inverse frequencies, the time component has been likened to a second-order Doppler shift that explains the gravitational red-shift . Whatever happened to the first-order shift? These superficial similarities are, indeed, unfortunate.

The conventional derivation of the orbital equations [2] begin with the metric:

$$ds^2 = -A(r)dt^2 + B(r)dr^2 + C(r)r^2 d\varphi^2, \tag{3.1}$$

---

[1] L D Landau & E M Lifshitz, *The Classical Theory of Fields* (Pergamon, Oxford, 1975) §101.
[2] Alder *et. al., op. cit.*

in the plane $\vartheta = \pi/2$ for which there is the Lagrangian:

$$\mathcal{L} = \frac{1}{2}\left(-A(r)\dot{t}^2 + B(r)\dot{r}^2 + C(r)r^2\dot{\varphi}^2\right), \tag{3.2}$$

where the dot denotes differentiation with respect to an affine parameter, and not with respect to proper time, because $ds = 0$ along null geodesics. [3]

The Lagrangian (3.2) gives the usual Euler-Lagrange equations:

$$\frac{d}{d\lambda}\frac{\partial\mathcal{L}}{\partial\dot{x}^\mu} - \frac{\partial\mathcal{L}}{\partial x^\mu} = 0.$$

It is clear from the Lagrangian, (3.2), that both t and φ are cyclic (aka ignorable, kinosthenic, etc.) in the sense that they are absent in the Lagrangian, and, yet, their derivatives are present. This is a wake-up call that there are two first integrals of the motion:

$$A(r)\dot{t} = \text{const} \equiv \ell \quad \text{and} \quad C(r)r^2\dot{\varphi} = \text{const.} \equiv L,$$

the former is associated with the conservation of energy, while the latter that of angular momentum.

The condition for characteristic surfaces in geometrical optics, to which rays are always orthogonal to, is the vanishing of the Lagrangian (3.2):

$$0 = -A\dot{t}^2 + B\dot{r}^2 + Cr^2\dot{\varphi}^2$$

which becomes

$$0 = -\frac{\ell^2}{A} + B\dot{r}^2 + \frac{L^2}{Cr^2},$$

when the first integrals are introduced. If the transform $d\lambda \to d\varphi$ is carried out, then the equation of the trajectory becomes

$$\frac{dr}{d\varphi} = \pm\frac{Cr^2}{L\sqrt{AB}}\sqrt{\ell^2 - \frac{L^2B}{Cr^2}}.$$

---

[3] The affine parameter, λ, must be chosen such that a null vector $\dot{x}^\mu$ preserves is length under parallel displacements. Employing this parameter with respect to the following variational principle determines the affine parameter up to a linear transformation.

Keplerian orbits will work just as well, but with $C = 1$ [4], although what these orbits
have to do with light trajectories is one of those coincidences based on the geodesic
nature of the motion.

3.1 Deflection of light

As we saw in § 2.1, conic sections result from the exact solution of these equations for
an inverse square Newtonian force. Small perturbations can also be handled by the
same method; a case in point is the deflection of light rays by a massive body. This
constitutes one of the three classic tests of GR, but it can be derived in a much more
intuitive, and correct way from geometrical optics.

The correction term that results in an angle of deflection has an inverse cube dependency whose numerical coefficient is the product of the angular momentum and
the Schwarzschild radius, seen in Eq (2.78). It is obtained by developing the $B =
1/(1 - \alpha/r)$ in a series in powers of the small coefficient, $\alpha \equiv 2\mu/c^2$.

In the absence of such a correction term, the equation of the trajectory is:

$$\varphi = \int_0^r \frac{\Delta dr}{r\sqrt{r^2 - \Delta}} = \cos^{-1}(\Delta/r),$$

where $\Delta = L/c$ is the impact parameter . But, this is an equation of a straight line,
$r = \Delta/\cos\varphi$, with no central mass to deflect from. So why should a perturbation give
the correct deflection of a ray of light about a massive body like the Sun?

The mass comes from the 'perturbation' itself: the Schwarzschild coefficient, $B =
1/(1 - \alpha/r)$ which converts the centrifugal potential, $V(r) = L^2/2r$, into:

$$V' = \frac{L^2}{2r}\left(1 - \frac{\alpha}{r}\right). \tag{3.3}$$

Both potentials have the gravitational term, $-GM/r$, ostensibly absent, and this explains why Einstein's result,

$$\varphi = 2\alpha/R_S, \tag{3.4}$$

---

[4]R A Crudo & J C O'Brien. "A Metric Approach to Transformation Optics"

where R_S is the radius of the Sun, cannot be considered as a 'modification' of $2\alpha/\Delta$. The latter leads to the Rutherford expression,

$$\cot(\varphi/2) = \Delta/\alpha,$$

for small angle deflections, and the gravitational parameter, μ, stands in for the product of the atomic number and charge. The speed of light corresponds to the speed of the deflecting particle.

The deflection occurs for a hyperbolic orbit of eccentricity,

$$\varepsilon = \sqrt{1 + \left(\frac{\Delta}{\alpha}\right)^2} > 1.$$

In Einstein's expression, (3.4), the radius of the Sun, R_S is confused with the distance of closest approach, that is, the impact parameter,

$$R_S = \alpha(\varepsilon - 1) \approx \frac{\Delta^2}{2\alpha}.$$

If the radius of the Sun were direRutherford.jpegth the distance of approach to a point target we would get a totally different result, as we now show.

In the Rutherford calculation, the target is a point target, as shown in Fig (3.1). Thus, if we wanted to replace the radius of the Sun with the impact parameter we would come out with a deflection of $2(\alpha/\Delta)^2$, which is the second-order term in α that Einstein neglected.

To make matters worse, the angular momentum:

$$\frac{r^2\dot{\varphi}}{1 - \alpha/r} = L, \tag{3.5}$$

is not an integral of the motion in the Schwarzschild metric. [5]

However, in the 'weak-field' approximation, which is employed in the calculation of the advance of the perihelion of Mercury, L is treated as a constant of the motion.

---

[5]C Moller, *The Theory of Relativity* (Clarendon, Oxford, 1952) p 349 Eq (18).

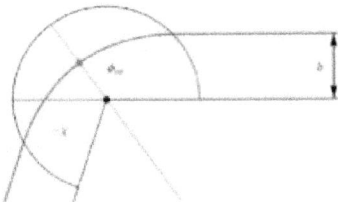

Figure 3.1: A trajectory coming in from infinity with impact parameter, Δ. The trajectory
approaches it closest distance to the point source at the radial turning point, indi-
cated by a dot. After which, the trajectory proceeds back to infinity, although in a
different direction. At the turning point the trajectory has deviated from a straight
line by the angle ϕ_∞. The total deviation will be twice as great. The scattering angle χ gives the deviation of the trajectory from its direction of incidence; it is the
difference between π and the total angle $2\phi_\infty$, as given in Eq (4.29) below.

This means that terms of order α/r — the same order as the deflection of light, and the
advance of the perihelion— have been neglected in order to render $r^2\dot\varphi$ a constant of
the motion.

In fact, Moller [6] claims that the left-hand side of (3.5)

> cannot be interpreted as angular momentum, since the notion of a 'ra-
> dius vector' occurring in the definition of the angular momentum has an
> unambiguous meaning only in Euclidean space.

By this, he intends to throw out the entire calculation of the perhelion advance and
the deflection of light because these are of the same order of magnitude as the term he
would have us neglect in (3.5).

Multiplying Moller's expression for L by the angular speed, $\dot\varphi$, but considering L as a
conserved quantity, there results:

$$\frac{r^2\dot\varphi^2}{1-\alpha/r} \simeq \frac{L^2}{r^2}\left(1+\frac{\alpha}{r}\right).$$

---

[6] *op. cit.* p 350.

Except for a sign, the right-hand side is the Einstein potential, (3.3) in Fig (2.26) so throwing out the correction term in (3.5) is tantamount to eradicating the term involved in both the deflection of light, and the perihelion advance of Mercury. The sole difference between the two is that the Newtonian potential is included in the latter.

3.2 Problems with the Conventional Formulation

The conventional derivation suffers from two inaccuracies:

■ the constraint $A\dot{t} = \ell$ cannot be introduced more than once into the variational equations, for, otherwise, what was minimum would now become a maximum, or *vice verse*, and

■ the condition for a first integral,

$$\frac{d}{d\lambda}\left(A\frac{dt}{d\lambda}\right) = 0,$$

gives $Adt/d\lambda = \ell$, and not $\dot{t} =$ const. because it is supposed that $A \neq A(\lambda)$. However, $A = A(r)$, and certainly $r = r(\lambda)$ so that the constraint is $A\dot{t} = \ell$ and not $\dot{t}$ alone.

The latter has an important consequence, because it violates the so-called condition of a *static* field: the components of the Ricci tensor like $R_{01} = -\dot{A}/Ar$ no longer vanish.

In view of the generalized Minkowski line element, the index of refraction can be written as

$$n^2(r) = B\dot{t}^2.$$

The similarity between coordinate and proper times is striking— but, also, extremely misleading because explicit time dependencies do not enter into GO. Any difference in the affine parameters will show up as wavelength changes – not as frequency changes. Time considerations do, however, enter into the rates of change of the coordinates, $(r, \vartheta, \varphi, t)$.

The lagrangian (3.2) can be written as:

$$\mathcal{L} = \frac{1}{2}\left(B\dot{r}^2 + r^2\dot{\varphi}^2 - n^2(r)\right), \tag{3.6}$$

where $n(r)$ is the index of refraction. It is to be considered as a constraint on the motion, just like ℓ was.

To determine n, consider the Legendre transform of the kinetic energy: $T = (1/2)B\dot{r}^2$,

$$U(\dot{t}) = \dot{r}\frac{\partial T}{\partial \dot{r}} - T,$$

where a new variable, $\dot{t}$ is

$$\dot{t} = \frac{\partial T}{\partial \dot{r}} = B\dot{r}.$$

Consequently, the new potential is:

$$U(\dot{t}) = \dot{t}^2/2B.$$

Thus, the Lagrangian becomes:

$$\mathcal{L} = \frac{1}{2}\left(B\dot{r}^2 + r^2\dot{\varphi}^2 - B^{-1}\dot{t}^2\right).$$

Then a comparison with our original Lagrangian shows that $n^2 = A\dot{t}^2$, and $AB = 1$, just as in the Schwarzschild metric.

Now, the Legendre transform defines $\dot{t} = B\dot{r}$, whereas the first integral of the motion gives: $A\dot{t} = \ell$. The only limiting speed is c, and, hence, $\ell = c$. Consequently, the Euler-Lagrange equation,

$$\ddot{r} + \frac{1}{2}B'\dot{r}^2 - r\dot{\varphi}^2 - \frac{B'}{2B^2}\dot{t}^2 = 0,$$

where the prime means differentiation with respect to r, becomes

$$\ddot{r} - r\dot{\varphi}^2 = -\frac{1}{2}B'(\dot{r}^2 - c^2).$$

Since the radial acceleration must be negative, $B' < 0$.

As an illustration, consider the Schwarzschild metric, where $B = (1 - \alpha/r)$, the radial equation of motion is

$$\ddot{r} - r\dot{\varphi}^2 = \frac{\alpha}{2r^2} \frac{\dot{r}^2 - c^2}{(1 - \alpha/r)^2}.$$

In the non-relativisitic limit, it reduces simply to

$$\ddot{r} - r\dot{\varphi}^2 = -g,$$

where $g = \mu/r^2$, Newton's law [7] Secondly, consider the Weber force, where $B = 1/r$. Since it is a radial force, $r\dot{\varphi}^2$ vanishes, and the Euler-Lagrange equation with an external force is precisely:

$$\ddot{r} - \frac{1}{2}\frac{\dot{r}^2}{r^2} + \frac{1}{2}\frac{c^2}{r^2} = F_W,$$

where c is the Weber constant, the $\sqrt{2}$ greater than the speed of light, i.e. $\ell = \sqrt{2}c$ (cf. Eq (2.14)).

The analogies are, however, only formal since λ is an affine parameter that measures the advance along the trajectory. Yet, it goes to show how similarities can often time be very deceiving.

The index of refraction has its feet in two worlds: one foot in the *static* world of light propagation through an inhomogeneous medium, and the other foot in the *kinetic* world because of its relation to the speed of propagation, $v = c/n$.

The geometric meaning of the index of refraction can be gleaned by considering trajectories of light rays in 'virtual' and 'real' space, or those on a sphere and on a plane, as shown in Fig. (2.16).

When virtual space is mapped onto real space the affine parameter, λ' will change into λ in the same interval of time. If light travels at speed c in virtual space, it will travel

---

[7]q.v. C H McGruder, III, "Gravitational repulsion in the Schwarzschild field," *Phys Rev* **D25** (1982) 3191, who comes out with repulsive terms.

at $(d\lambda/d\lambda')c$ in real space, or $d\lambda/dt = c/n$, where

$$n = \frac{d\lambda'}{d\lambda}, \tag{3.7}$$

the ratio between the increments $d\lambda'$ and $d\lambda$. When the trajectory of a light ray moves $d\lambda$ in real space, it will move $d\lambda'$ in virtual space.

So this is the equivalence between the kinetic definition $d\lambda/dt = c/n$ and the static definition, 3.7).

To see this in more detail, consider the case of the Schwarzschild field, $A = 1 - \alpha/r$, so that there will be a shift in the affine parameter by an amount:

$$\lambda' = (1 - \alpha/r)^{-1/2}\lambda.$$

Therefore, in a medium of varying index of refraction, the affine parameter, which can be measured in the number of wavelengths of light that is used, will vary. The wavelength will change because the leading edge of the wave will encounter a new value of the index of refraction before the trailing edge does. In this case the wavelength and velocity of the wave will change but not the frequency. This is what we expect will happen when light propagates in an inhomogeneous gravitational field. In other words, gravitational fields do not cause red-shifts.

This is the antithesis of what GR professes [8]. Motion causes frequencies to vary, as Christian Doppler discovered, while light propagating in a static gravitational field can only vary its wavelength in response to changes in the speed of propagation, as Snell discovered. In other words, "clocks do not slow down in gravitational fields, but the speed of light does because of changes in wavelength."

Hence, our interpretation of the gravitational metric,

$$ds^2 = g_{\nu\mu}dx^\nu dx^\mu,$$

is in complete agreement with Levi-Civita [9] In addition to the vanishing of the Lagrangian, (3.2), wave surfaces will also be required to satisfy the Hamilton-Jacobi equa-

[8]q.v. C Moller, *op. cit.* §129.
[9]T. Levi-Civita, *Rend Acc Lincea* **11** (1930) 3 & 113.

tion:

$$H = \frac{1}{2} g^{\nu\mu} p_\nu p_\mu = 0,$$

where the p^μ are the conjugate momenta. As we mentioned in the introduction to this chapter, even Landau & Lifshitz [10] find that this is the only way to introduce the peripheral mass m, for which the Hamilton-Jacobi equation is $H - m^2c^2 = 0$.

However, for light the characteristics satisfy equations of first-order, $H = 0$. Light propagates along null geodesics that are defined by the indefinite metric, ds^2. He ends his discussion by admitting that the theory of the characteristic of GR is not new, and was clearly formulated in a note by Edmund Whittaker [11] entitled, "Note on the law that light-rays are the null geodesics of the gravitational field. Both authors are, therefore, in complete agreement that the gravitational potentials, $g_{\nu\mu}$, describe electromagnetic disturbances, and not gravitational radiation.

3.3 Gravitational repulsion in the Schwarzschild field

The title of this section is taken from McGruder' s article [12], and it is a classic example of deductions from inaccurate suppositions.

The whole idea of introducing and affine parameter, λ, is that we want the Lagrangian to be valid along geodesics where the proper time interval, $d\tau = 0$. So we are talking about trajectories of light rays, and it makes no sense to talk about *repulsive* gravity! There is no 'test' particle to which we can attribute the attribute of repulsive gravity.

The Lagrangian for the Schwarschild metric is:

$$\mathcal{L} = \frac{1}{2} \left(-Bc^2 t'^2 + Ar'^2 + r^2 \varphi'^2 \right), \tag{3.8}$$

---

[10]L D Landau & E M Lifshitz, *Classical Theory of Fields* (Pergamon Press, Oxford, 1975) §101.
[11]E Whittaker, *Proc Camb Phil Soc* **XXIV** (1927) 32.
[12]*op. cit.*

in the plane, $\varphi = \pi/2$, where

$$B = A^{-1} = (1 - \alpha/r).$$

The prime stands for differentiation with respect to the affine parameter, λ.

The Lagrangian, (3.8), comes with two integrals of the motion,

$$Bt' = H, \qquad \text{and} \qquad r^2\varphi' = L.$$

The condition for null geodesics is the vanishing of the Lagrangian, the condition; it
is:

$$Ar'^2 = \frac{(cH)^2}{B} - \frac{L^2}{r}. \tag{3.9}$$

As a preliminary, let us change the differentiation variable to φ using:

$$r' = \frac{dr}{d\varphi}\frac{dt}{d\lambda} = \frac{L^2}{r^2}\frac{dr}{d\varphi}.$$

When we insert this into (3.9) we come out with:

$$\frac{A}{r^4}\left(\frac{dr}{d\varphi}\right)^2 = \frac{c^2 H^2}{B L^2} - \frac{1}{r^2}. \tag{3.10}$$

The important point to observe is *the values of the first-integrals must always appear in
a ratio, and not separately.* Dividing through by A, using $AB = 1$, and changing the
integration variable to $u = 1/r$ give:

$$\left(\frac{du}{d\varphi}\right)^2 = \left(\frac{cH}{L}\right)^2 - u^2(1 - \alpha u).$$

We should be happy with the first-order equation because we need a single integration
constant. But, we aren't happy, and go ahead and differentiate a second time getting:

$$\frac{d^2u}{d\varphi^2} + u = \frac{3}{2}\alpha u^2,$$

which is Einstein's modification of the equation of the orbit that we found earlier in
(2.78). This complicates life in that we need another constant of integration.

Recall what Danby [13] has to say about it:

> In general (and as long as we have a conservative field),
>
> $$\frac{d^2u}{d\varphi^2} + u = \frac{1}{L^2u^2}f\left(\frac{1}{u}\right), \qquad (3.11)$$
>
> is not so useful as [(3.10)] since the latter is only [sic] a first-order differential equation. In fact, [(3.11)] can be recovered (rather laboriously) by differentiating [(3.10)], and its only advantage occurs when $f(1/u)$ has such a form that the solution can be written down directly; this is the case when
>
> $$f\left(\frac{1}{u}\right) = au^2 + bu^3.$$

Having shown that the transformation from the affine variable to φ gives the Einstein modification of the orbit, (2.78), we are now ready to take on *repulsive* gravity, by transforming the affine parameter to t.

Introducing:

$$r' = \dot{r}t' = \dot{r}\frac{H}{B},$$

into (3.9) we get:

$$\frac{\dot{r}^2}{(1-\alpha/r)^3} + \left(\frac{L}{rH}\right)^2 = \frac{(cH)^2}{1-\alpha/r}. \qquad (3.12)$$

When we differentiate (3.12) with respect to time, we find:

$$\ddot{r} - r\dot{\varphi}^2/H = g\left\{\frac{3\dot{r}^2}{1-\alpha/r} - (1-\alpha/r)\right\}, \qquad (3.13)$$

which is exactly what Treder & Fritze [14] find for radial acceleration, realizing that we are free to set $H = 1$.

Next, McGruder claims that: (3.13) reduces to the Newtonian expression, $\ddot{r} = -g < 0$, "for small particle velocities ($\dot{r} \ll c$) [and] in weak fields ($r \gg \alpha$)." But, what 'particles' might these be, and weak fields of what?

[13] J M A Danby, *Fundamentals of Celestial Mechanics* (MacMillan, New York, 1962) p 61.
[14] H J Treder & K Fritze, *Astron Nachr* **296** (1975) 109.

Continuing McGruder says, in Einstein's theory, however, Eq (3.13), shows that for

$$\dot{r}^2 > \frac{c^2}{3}(1 - \alpha/r)^2,$$

$\ddot{r} > 0$, "implying gravitational repulsion."

However, since we are dealing with conservative fields we can deal with (3.12) directly
and find:

$$\dot{r}^2 - c^2(1 - \alpha/r)^2 = -\frac{L^2}{r^2}(1 - \alpha/r)^3. \tag{3.14}$$

Expanding $(1 - \alpha/r)$ in powers of α, and retaining linear terms, result in:

$$\dot{r}^2 + 4\frac{\mu}{r} + \frac{L^2}{r^2}(1 - 3\alpha/r) \simeq c^2. \tag{3.15}$$

A comparison with what is referred as the conservation of energy:

$$\dot{r}^2 + 2V_E = c^2, \tag{3.16}$$

where:

$$V_E = -\frac{\mu}{r} + \frac{L^2}{r^2}(1 - \alpha/r), \tag{3.17}$$

commonly referred to as the 'Einstein' potential, (cf. Fig (2.26)) [15] shows that gravity
is, indeed, repulsive — at least for 'weak' fields — and there is no reason why it should
become attractive for strong fields. In fact, the energy conservation equation, (3.16), is
an incompatible mixture of non-relativistic and ultra-relativistic terms.

---

[15]R Sexl & H Sexl, *White Dwarfs-Black Holes*, (Academic Press, New York, 1979), p 103.

4 The Optical Properties of Gravity

4.1 The Optics of Kepler's Laws

Kepler gives the orbit of a particle on an ellipse in time t as,

$$M = t\sqrt{\frac{\mu}{a^3}} = E - \varepsilon \sin E, \tag{4.1}$$

or, equivalently, in terms of the mean anomaly, M. a is the radius of a circle that inscribes the ellipse, as shown in Fig (4.1).

E is the central angle, given the sophisticated name of eccentric anomaly , and ε is the eccentricity.

Kepler's equation has played an important role in the history of mathematics. Beginning with Newton, the mean anomaly, M, which is transcendental, has been sought in a power series of the eccentricity, ε. It was found that the series converges for values of $\varepsilon < 0.6627\ldots$

Kepler's equation, (4.1), can be simply derived from the difference in the area of sector $a^2 E/2$ and the area of the triangle, $\triangle CfP$, namely $\frac{1}{2}a\varepsilon \times a \sin E$:

$$A = \frac{1}{2}a^2(E - \varepsilon \sin E).$$

Since the ellipse is a squashed circle, this area must be reduced by a factor of b/a, the ratio of the semi-minor to the semi-major axes of the ellipse.

Changing the sign of the eccentricity merely changes the position of the directrix from the right of the focus to the left of it. We will soon appreciate that ε plays the same role that the relative velocity plays in special relativity.

The angles v and E are related by:

$$\cos v = \frac{\cos E - \varepsilon}{1 - \varepsilon \cos E}$$

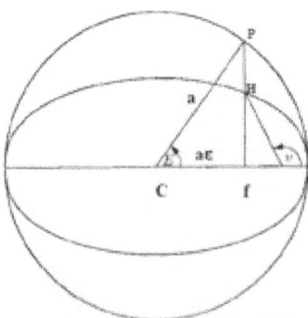

Figure 4.1: An ellipse can be considered as a locus of points such that the ratio of segments PH/Hf is always constant.

and

$$\sin v = \frac{\sqrt{(1 - \varepsilon^2)} \sin E}{1 - \varepsilon \cos E},$$

which bear a striking resemblance to aberration formulas where ε plays the role of a relative velocity.

We can drive this analogy home, by realizing that the two angles v and E are related by:

$$\tan \frac{v}{2} = \left(\frac{1 + \varepsilon}{1 - \varepsilon}\right)^{1/2} \tan \frac{E}{2}, \tag{4.2}$$

a familiar expression from aberration if ε is interpreted as the relative velocity between two inertial frames.

An analytic derivation of the mean anomaly (4.1) consists in assuming that the eccentric anomaly is defined by:

$$r = a(1 - \varepsilon \cos E). \tag{4.3}$$

Taking the time derivative of (4.3), and introducing it into the conservation of energy equation

$$n^2 r^2 - r^2 \dot{r}^2 = L^2, \tag{4.4}$$

where $L = b^2/a = a(1 - \varepsilon^2)$ and we have used Kepler's second.

Equation (4.4) is none other than the equation for the optical path of a ray. The time derivative replaces the derivative with respect to the angle v. In astronomical terms, n is referred to as the speed of the orbit ; however, we would prefer to identify it with the index of refraction of an Eaton lens.

It was observed by Luneburg [1] that light rays which propagate in a medium of index of refraction (2.48) are identical with the paths of particles that move in a Coulomb potential, $-1/r$, and an energy $-a/2$. This is rather odd, since we would normally expect that the index of refraction would vary inversely with the speed of the light ray. We will return to this shortly.

Depending on our reference focus, we have two mean anomalies, $t_\mp$ that are related by

$$\frac{dt_+}{dt_-} = n^2. \tag{4.5}$$

This clearly shows that time retardation is due to the propagation of light rays in a medium $n \neq 1$.

Introducing the definition of the eccentric anomaly into the expression for the index of refraction, (2.48), leads to

$$\frac{dt_+}{dt-} = \frac{1 + \varepsilon \cos E}{1 - \varepsilon \cos E}.$$

Differentiating (4.3) with respect to t, and introducing it into the conservation of energy equation, result in

$$\frac{a^3}{\mu} \dot{E}^2 (1 - \varepsilon \cos E)^2 = 1.$$

Since the orbit is described by $\dot{E} > 0$, taking the square root gives

$$dt = \sqrt{\left(\frac{a^3}{K}\right)} (1 - \varepsilon \cos E) dE.$$

Integration leads immediately to the mean anomaly, (4.3).

---

[1] R K Luneburg, *Mathematical Theory of Optics* (U California Press, 1966) p 169.

If t in that expression coincides with t_-, then (4.5) shows that

$$t_+ \sqrt{\left(\frac{\mu}{a^3} \right)} = E + \varepsilon \sin E. \qquad (4.6)$$

Since the right-hand sides of (4.1) and (4.6) are proportional to the areas, A_- and A_+, respectively, swept out in times t_- and t_+, the law of areas states that

$$\frac{A_-}{t_-} = \frac{A_+}{t_+} = \text{const.}$$

It is the same time transformation that we used in the previous chapter under the conformal transformation, $w = z^2$. The two times are related by $\varepsilon \to -\varepsilon$, analogous to a reversal in the direction of motion.

Expressing (4.5) in terms of the relative velocity u, we get

$$\varepsilon \cos E = \frac{1 - u^2}{1 + u^2},$$

which is the Doppler shift when light rays are reflected normally. Although our index of refraction, n, is the linear speed of the orbit in units of mean solar days [2], the transformation $\varepsilon \to -\varepsilon$ is tantamount to considering the relative velocity u as the inverse of the index of refraction, in line with its usual definition.

4.2 From Kepler's ellipse to the angle of parallelism

Kepler's laws are ordinarily thought of as being static with time entering only through the period of the motion. Yet, it may come as a surprise that time is involved, and, in fact, there are two different times depending on which focus is used for reference. Not only will we show the compatibility of Kepler's laws with geometrical optics, but, moreover, it will allows us to draw the analogy between the eccentricity and a constant relative velocity, something we touched on in the last section.

---

[2]P Herget, *The Computation of Orbits*, Eq (3.13).

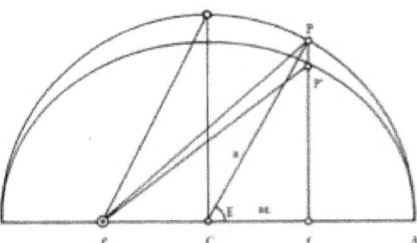

Figure 4.2: The eccentric circle inscribes the ellipse, which is defined by a locus of points P'
such that the ratio $P'f/Pf$ is constant. For the far focus f' the area fCP must
be added to the area of the sector $Ea^2/2$, instead of subtracting it as we would for
the near focus, f.

We may generalize the definition of the eccentric anomaly, (4.3), to:

$$r = a\left(1 \pm \varepsilon \cos E\right), \tag{4.7}$$

depending on which focus, f or f', we are considering in Fig (4.2).

Equating (4.7) with the usual equation of the ellipse,

$$1 \mp \varepsilon \cos E = \frac{(1 - \varepsilon^2)}{1 \pm \varepsilon \cos \theta}, \tag{4.8}$$

relates the eccentric anomaly, E, to the true anomaly, θ.

Solving (4.8) for the cosine and sign of the true anomaly gives the relations:

$$\cos \theta_\pm = \frac{1 \pm \varepsilon}{1 \pm \varepsilon \cos E}, \tag{4.9}$$

and

$$\sin \theta_\pm = \frac{\sqrt{1 - \varepsilon^2}\,\sin E}{1 \pm \cos E}. \tag{4.10}$$

As we mentioned in the last section (4.9) and (4.10) bear a striking resemblance to
aberration formula if the eccentricity, together with its sign, can be identified as a
relative velocity. Here we want to further develop the analogy.

For the *near* focus, f, the mean anomaly, (4.1), is defined as the *difference* in the sector area $Ea^2/2$ and the area of the triangle, CfP, which is $a^2 \sin E/2$:

$$\frac{M}{2\pi} = \frac{a^2(E - \varepsilon \sin E)}{2\pi a^2}.$$

For the *far* focus, f' instead of taking the difference of the areas, we take their *sum* and obtain $M = E + \varepsilon \sin E$.

We can write (4.10) as:

$$\sin \theta = \frac{b}{r} \sin E,$$

where $b = a\sqrt{1 - \varepsilon^2}$. Introducing $d\theta/dE = b/r$, yields:

$$\frac{d\theta}{\sin \theta} = \frac{dE}{\sin E}.$$

This can be integrated to obtain:

$$\ln \tan(\theta/2) = \ln \tan(E/2) + \text{const.}$$

The constant of integration may be identified through a comparison with (4.2) (where we now write θ for v):

$$\frac{1}{2} \ln \left(\frac{1 + \varepsilon}{1 - \varepsilon} \right) = \tanh^{-1} \varepsilon = \bar{\varepsilon},$$

where $\bar{\varepsilon}$ is the *hyperbolic* measure of the eccentricity. The solution can thus be written as:

$$\tan(\theta/2) = e^{\bar{\varepsilon}} \tan(E/2).$$

Setting $\theta = \pi/2$, which is the angle that Pf makes with Af', so that $P'f$ is identified as the semi-latus rectum, we obtain:

$$\tan \frac{E}{2} = \exp(-\bar{\varepsilon}). \tag{4.11}$$

Equation (4.11) defines the angle of parallelism, E.

By introducing a new variable,

$$\gamma = \frac{2\varepsilon}{1 + \varepsilon^2}, \tag{4.12}$$

we can make the connection with dual laws. First, let us remark that (4.12) transforms the Poincaré disc motion into the Klein model. Both are models of the hyperbolic plane, but whereas the Poincaré model is conformal (it preserves angles), the Klein model does not.

The isomorphism takes straight 'lines,' or chords, in the Klein model and bends them into circular arcs that cut the rim orthogonally in the Poincaré model. If ε is the distance from the center of disc to any point P, then the rays $\vec{OP} = \vec{OK}$, where K is the corresponding point in the Klein model. Their, magnitudes, $|OK| = \gamma$ and $|OP| = \varepsilon$, will be given by (4.12). Alternatively, if we have two bodies traveling at equal and opposite speeds $v = |v_1| = |v_2| = \varepsilon$, their relative speed will be given by $v_{12} = \gamma$,

$$v_{12} = \frac{2v}{1 + v^2}.$$

Second, if we set $w = 1/\gamma$ and $z = 1/\varepsilon$, (4.12) becomes none other than the Joukowski transform:

$$w = \frac{1}{2}\left(z + \frac{1}{z}\right),$$

which transforms circles into ellipses with the same centers. We will also see that it transforms chords of a circle into circular arcs that cut a disc orthogonally.

Treating w and z as complex variables, we have that when the point Z describes a circle of radius $|z| = r > 1$, the point $w = z + 1/z$ describes an ellipse with its center at the origin. Squaring an ellipse gives another ellipse, but with one of its foci transferred to the origin! We have thus gone from a Hookean ellipse to a Newtonian ellipse, as we have discussed §2.2.1.

As a final point, we know that $|z| = r$ is a circle, and $|z \pm f| = \text{const.}$ is an ellipse. Their sum is also constant,

$$|z + f| + |z + f'| = \text{const.}$$

In fact, this is the definition of an ellipse: The sum of the distances of each point from the two foci, f and f' is constant. This is related to the arithmetic mean being constant. But what if the geometric mean, or the product of the distances,

$$|z + f| \cdot |z + f'|,$$

were to be constant? The first person who thought about such a possibility was Giovanni Cassini, and in 1680 he initiated his study of what were to be known as Cassini ovals. For Cassini ovals, it is the product of the distances from two fixed foci which is constant. We will come back to our discussion of Cassini's ovals in §4.5.

4.3 Snell's law in rotating systems

We are accustomed to the fact that the index of refraction is *always* inversely proportional to the velocity. Yet, for rotational motion, the precise orbital speed of a body at any given point in its trajectory is:

$$v = \sqrt{\mu \left(\frac{2}{r} - \frac{1}{a} \right)}, \tag{4.13}$$

where μ is the standard gravitation parameter, r is the distance at which the speed is to be determined, and a is the length of the semi-major axis of the ellipse.

Expression (4.13) is referred to as the *vis-viva* equation. [3] We can appreciate that the term in the parenthesis in (4.13) is precisely the square of the Eaton index of refraction, (2.48). As such, the *vis-viva* envisions an index of refraction that is proportional to the velocity – and not inversely proportional to it.

This recalls the debate between the Minkowski and Abraham debate about which form of the momentum is correct for relativistic particles. According to Minkowski, the correct form of the momentum is $p = nE/c$, and since the index of refraction is $n = c/v$, the relativistic energy of a free particle is $E = pv$. Alternatively, if we listen to Abraham, the momentum is $p = E/nc$, and for the same index of refraction the energy is $E = (p/v)c^2$.

---

[3]T Logsdon, *Orbital Mechanics: Theory & Applications* (Wiley, New York, 1998).

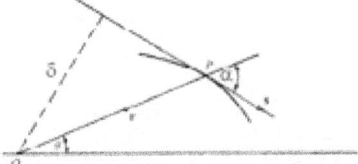

Figure 4.3: Illustration of Bouguer's formula showing that $n\delta = $ const. in a medium of index of refraction n possessing spherical symmetry.

Now look what happens if we take the index of refraction proportional to the velocity, $n = v/c$, making it less than unity. In the case Minkowski, the momentum is $p = vE/c^2$, while, in the case of Abraham, $p = E/v$. Minkowski and Abraham have swapped expressions!

Who is right? The last time we checked, the jury is still out for deliberation. [4] We begin with the orbital equation of the trajectory [cf. Eq (4.24) below]

$$\varphi' \equiv \frac{d\varphi}{dr} = \pm \frac{L}{r\sqrt{(nr)^2 - L^2}}$$

Resolving for the angular momentum, L, we find:

$$L = \pm \frac{r^2 \varphi'}{\sqrt{1 + r^2 \varphi'^2}} = \pm nr \sin \alpha, \qquad (4.14)$$

where α is the angle the ray makes with the tangent to the ray at the point P in Fig (4.14).

Equation (4.14) is commonly referred to as Bouguer's formula. . Born & Wolf [5] refer to it as "the analogue of a well-known formula in dynamics, which expresses the conservation of angular momentum of a particle moving under a central force."

However, Newton told us earlier that this law is given by (2.2), v is the tangential velocity to the ray at the point P. And since $\sin \alpha = \delta/r$, which is the perpendicular

---

[4]M Buchanan, *Nature Physics* **3** (2007) 73.

[5]M Born & E Wolf, *Principles of Optics* (Pergamon Press, Oxford, 1959) p 122.

distance from the tangent of the ray to the origin at O, the product $n\delta$ = const. if the angular momentum is to be conserved.

Now, if follows from (2.2), and Bouguer, (4.14), that $n = v$ in units where $c = 1$. There are numerous of papers in the literature that draw attention of Newton's second law, $F = ma$, in optics, but with the index of refraction always proportional to the inverse of the velocity. They claim that geometrical optics is the 'zero-energy' analogue of mechanical orbits, going so far as to write $L = n^2 r^2 \dot{\varphi}$. [6]

We recall that Newton also defined the slope z as: $z = 1/r\varphi' = \cot\alpha$, so that

$$1 + z^2 = \left(\frac{nr}{L}\right)^2 - 1 = \csc^2\alpha.$$

This leads immediately to the relation,

$$nr\sin\alpha = n_0 r_0 \sin\alpha_0 = L,$$

where r_0 and α_0 are integration constants that determine the origin of the ray, the constant L, and its direction at P.

Setting $r = r_0$ = const., we get what looks like Snell's law [cf. Eq (5.26) below]:

$$n\sin\alpha = n_0 \sin\alpha_0.$$

It is usually the case that when light passes from a medium of higher index of refraction to one of lower index of refraction, the wavelength increases because the leading edge of the wave enters first and speeds up, while the trailing edge lags behind. The wave begins to stretch out so that both the wavelength and its speed change, but the frequency, ν, remains constant.

Here, no such thing happens because the index of refraction is equal to the angular velocity, $n = r\omega$, where $\omega = 2\pi\nu$, the angular speed so that under the same condition that $r = r_0$ = const., Snell's law becomes :

$$\omega\sin\alpha = \omega_0 \sin\alpha_0. \tag{4.15}$$

---

[6] J Evans & M Rosenquist, "'$F = ma$' optics,", textitAm J Phys **54** (1986) 876-883.

Rotational motion gives the impression of a temporal variation where both the angular speed and speed of light change, but the wavelength does not. Because the leading edge of the wave never encounters a new value of n before the trailing edge does, the wavelength will not change. The change in angular speed therefore entails a change in the speed of light.

As an illustration of the equivalence between the index of refraction and the angular velocity, we can derive Newton's inverse square law without any recourse to Kepler's areal law. Solving Eaton's index of refraction, (2.48) for the angular velocity gives:

$$r\omega = \sqrt{\frac{2a - r}{ra}}.$$

Recalling that the force directed to the center is first-order aberrated, (2.3), where the radius of curvature is now given by:

$$\varrho = \frac{(2a - r)^{3/2} r^{3/2}}{a^{3/2} L}.$$

An advantage of this is that the angular momentum will cancel out in the final result so we don't have to specify that it is conserved.

We can write Bouguer's formula (4.14) in the form:

$$\sin\alpha = L/\sqrt{r(2a - r)/a},$$

and when these expressions are introduced into (2.3), we come out precisely with Newton's inverse square law, $F_0 = 1/r^2$.

4.4 The relativity of Kepler's laws

In the section on dual laws, we discussed *repulsive* gravity within the context of a conformal equivalence between the square of Hookean ellipses and the orbits in a gravitational field. We now want to show that the relation between Hookean and Newtonian forces lead to non-Euclidean geometries.

Milnor [7], in order to connect Keplerian orbits with the disc model of non-Euclidean geometries, used Hamiltons principle that the velocity vector, $\vec{v}$, of any non-degenerate Keplerian orbit, moves along a circle. What he shows is that given the total energy, E, then a manifold of all vectors $\vec{v}$ such that $\vec{v} \cdot \vec{v} > 2E$ possesses a Riemann metric of constant curvature $K = -2E$. The geodesics on this manifold consist of circles that are associated with unique Keplerian orbits. However, in order to do so, Milnor had to resort to inverse velocities, which eradicates the limiting speeds of the orbits that are dictated by special relativity.

Begin with Newton's differential equation:

$$\frac{d^2\vec{x}}{dt^2} = -\mu\vec{x}/|\vec{x}|^3.$$

Levi-Civita recognized that it is possible to simplify solutions to this equation by introducing a 'fictitious' time parameter $s = \int dt/r$, where $r = |\vec{x}|$. Then writing Newton's law as $|d\vec{v}/dt| = \mu/r^2$, and dividing it by $ds = dt/r$, result in

$$\left|\frac{d\vec{v}}{ds}\right| = \vec{v} \cdot \vec{v}/2 - E,$$

on the strength of the energy conservation law:

$$\frac{\mu}{r} = \vec{v} \cdot \vec{v}/2 - E.$$

Inverting this relation and squaring lead to the line element:

$$ds^2 = \frac{4d\vec{v} \cdot d\vec{v}}{(\vec{v} \cdot \vec{v} - 2E)^2}.$$

This still does not look like a Riemann metric, so Milnor introduced inverse velocities, $\vec{u} = \vec{v}/|\vec{v}|^2$. Thus, $d\vec{u} \cdot d\vec{u} = d\vec{v} \cdot d\vec{v}/(\vec{v} \cdot \vec{v})^2$, so that the line element becomes precisely the form of a Riemann metric,

$$ds^2 = \frac{4d\vec{u} \cdot d\vec{u}}{(1 - K\vec{u} \cdot \vec{u})^2}, \tag{4.16}$$

[7]J Milnor, "On the geometry of the Kepler problem," *Am Math Mon* **90** (1983) 353-365.

with constant curvature, $K = 2E$. In this way, the two non-Euclidean geometries of constant curvature are accounted for: $K > 0$ hyperbolic, and $K < 0$, elliptic geometries.

However appealing the derivation is, we would like to retain the limits of velocity, and avail ourselves of things like velocity composition laws, and c as the limiting speed. As a bonus, we will appreciate that what are usually considered relativist phenomena, like time dilation and space contraction, also belong to conformal mappings.

Consider the conformal mapping, $w = z^2$. We would like to believe that Kepler's areal laws are invariant under such transforms, viz.,

$$\frac{|z|^2\dot{\varphi}}{dt} = \frac{|w|^2\dot{\varphi}}{d\tau} = L = \text{const.}$$

The two times are related by $d\tau = |z|^2 dt$, and if z is a solution to the harmonic oscillator differential equation:

$$\frac{d^2 z}{dt} = -z,$$

it will have a total energy of

$$|\dot{z}|^2 + |z|^2 = 2E.$$

Now, solving for $|z|$, and introducing it into the expression for $d\tau$ lead to the conclusion that the time, τ, undergoes time dilation:

$$d\tau = \left(2E - |\dot{z}|^2\right) dt.$$

Likewise, space contraction follows from taking the derivative of the conformal mapping, and taking its absolute value to obtain:

$$\frac{1}{2}\left|\frac{dw}{dz}\right| = |z| = \sqrt{\left(2E - |\dot{z}|^2\right)}.$$

If we take the absolute value of the harmonic oscillator equation,

$$\left|\frac{d^2 z}{dt}\right| = \sqrt{\left(2E - |\dot{z}|^2\right)},$$

rearrange it, and then square it, we come out with:

$$dt^2 = \frac{d\dot{z} \cdot d\dot{z}}{2E - |\dot{z}|^2}. \tag{4.17}$$

This still does not look like a Riemann metric, (4.16) for we should expect the denominator to be squared.

We know that the velocity metric is: [8]

$$dt^2 = \frac{2E|d\dot{z}|^2 - |\dot{z} \wedge d\dot{z}|^2}{(2E - |\dot{z}|^2)^2}, \tag{4.18}$$

where the last terms takes into account that the velocity vectors may not lie in the same plane. Writing this term as: of (4.18) as:

$$|\dot{z} \wedge d\dot{z}|^2 = |\dot{z}|^2|d\dot{z}|^2 - |\dot{z} \cdot d\dot{z}|^2,$$

and observing that for *uniform* acceleration , the increment in the velocity, $d\dot{z}$, is always perpendicular to the tangential velocity, $\dot{z}$, so that the last term vanishes. Consequently, (4.18) reduces to (4.17).

This is Newton's initial assumption where the centripetal acceleration points inward toward the center of the circular orbit. If we denote the velocity by $\vec{v}$, the radius of curvature is:

$$\varrho = \frac{|\vec{v}|^3}{|\vec{v} \wedge \dot{\vec{v}}|}.$$

And for uniform acceleration, the denominator reduces to $|\vec{v}||\dot{\vec{v}}|$, and, consequently, the radius of curvature simplifies to:

$$|\dot{\vec{v}}| = \frac{|\vec{v}|^2}{\varrho},$$

---

[8]B. H. Lavenda, *A New Perspective on Relativity: An odyssey in non-Euclidean geometries* (World Scientific, Singapore, 2011), Ch. 2.

which we recall is Newton's point of departure, (2.3), in his proof of his inverse square law.

So it is as Newton claimed it to be: Given the curvature that the force inflicts on the orbit, the path is uniquely determined. All inverse square laws lead to conic sections depending on the sign of the total energy. Their geometries are described by Riemann metrics of constant curvature. In this way we avoid introducing inverse velocities like Milnor did.

What can we say about the line element of its confomal dual, w? To this end we employ the Maupertuis-Jacobi principle

$$E - V(z) = \left| \frac{dw}{dz} \right|^2 (E' - V'(w)), \qquad (4.19)$$

which relates the potential, V, and total energy, E, of z to the corresponding quantities, V' and E' of its dual, w. The pairs of energies are required to satisfy:

$$EE' = V(z)V'(w) = \text{a negative const.}$$

Consider the attractive scenario. The oscillator has a potential energy, $V(z) = |z|^2$. According to the Maupertuis-Jacobi principle, this requires a negative total energy, $E' = -1/4$ on the right-hand side of (4.19). And a constant total oscillator energy of $E = 1$, requires an attractive gravitational energy of

$$V' = -\frac{\mu}{4|w|},$$

where μ is a positive constant, whose numerical value is of no importance.

Applying the law of conservation of energy to both oscillator and gravitational fields leads to

$$\frac{1}{2}|\dot{z}|^2 = E - |\dot{z}|^2,$$

and

$$\frac{1}{2}|\dot{w}|^2 = \frac{\mu}{|w|} - |E'|,$$

respectively. The absolute sign is necessary in the last term since $E' < 0$.

Newton's law of gravitation is: [9]

$$\frac{d^2w}{d\tau^2} = -\frac{\mu w}{|w|^3}.$$

Taking the absolute value of both sides of Newton's law,

$$\left|\frac{d\dot{w}}{d\tau}\right| = \frac{\mu}{|w|^2},$$

and changing times,

$$|d\dot{w}| = \frac{\mu dt}{|w|} = \left(|E'| + |\dot{w}|^2/2\right) dt,$$

give the line element

$$dt^2 = \frac{4d\dot{w} \cdot d\dot{w}}{(2|E'| + |\dot{w}|^2)^2}.$$

But, this is none other than the stereographic line element of elliptic geometry (q.v., Fig (2.16) and the discussion therein).

We recall that Milnor [10] resorts to the introduction of *inverse* velocities to describe what is happening "in the neighborhood of infinity." However, such a 'neighborhood' does not exist for velocities, like it does for the radial coordinate. It is therefore not always true that "$\vec{v} \cdot \vec{v} > 2E'$ must always be satisfied."

For *repulsive* gravity we get the hyperbolic line element:

$$dt^2 = \frac{4d\dot{w} \cdot d\dot{w}}{(2E' - |\dot{w}|^2)^2}.$$

This is none other than the Poincaré disc model where $|\dot{w}|^2/2E' < 1$, $2E'$ being the rim of the velocity disc.

---

[9] Arnol'd, *op. cit.* and other authors introduce a constant on the right-hand side of the equation which can be both positive (implying elliptic orbits), and negative (implying hyperbolic orbits). Their constant is related to the total energy E', but it is not the energy that changes sign, rather it is the coefficient. In our opinion the introduction of such a constant camouflages the physics.

[10] *op. cit.*

Translated back into oscillator terms, the metric is:

$$dt^2 = \frac{4d\dot{z} \cdot d\dot{z}}{2|E| - |\dot{z}|^2}. \tag{4.20}$$

Hitherto, these metrics were derived from the hyperbolic law of addition of velocities, [11] yet follow directly from 'velocity' circles of Keplerian orbits.

Consequently, the Riemann metric:

$$dt^2 = \frac{4d\vec{v} \cdot d\vec{v}}{(1 - K\vec{v} \cdot \vec{v})^2},$$

of constant curvature, K is not restricted to relativistic phenomena where the velocities obey the hyperbolic law of addition, for $K > 0$. This includes also the metric (4.20), which applies to uniform acceleration.

4.5 From Kepler ellipses to Cassini ovals

In this section we want to investigate more fully the relation between an optical medium with an index of refraction,

$$n = 1/(1 + r^2),$$

and one with an index of refraction

$$n^2 = 1/r + C,$$

where C is a constant of integration. [12] The former describes the closed optical paths of Maxwell's fish-eye on a sphere, while the latter the Keplerian ellipses in the plane.

The equation of wavefronts, S, in the Keplerian case is: [13]

$$(\nabla S)^2 = C + \frac{1}{r}. \tag{4.21}$$

---

[11]V Fock, *The Theory of Space Time and Gravitation* (Pergamon Press, New York, 1959) Ch 1.
[12]q.v., Eq (2.48
[13]Luneburg, *op. cit.*, p 172ff.

Now with $\nabla S = \vec{z}$ and $\vec{r} = \nabla W$, such that the Legrendre transform is

$$S + W = \vec{r} \cdot \vec{z}, \tag{4.22}$$

the Hamilton-Jacobi (4.21) becomes:

$$\vec{z} \cdot \vec{z} = C + \frac{1}{\sqrt{\nabla W \cdot \nabla W}}.$$

Rearranging and squaring give the wave fronts for Maxwell's fish-eye:

$$(\nabla W)^2 = \frac{1}{(-C + \vec{z} \cdot \vec{z})^2}.$$

For $C = -1$, the wave fronts are those with an index of refraction:

$$n = \frac{1}{1 + \vec{z} \cdot \vec{z}}. \tag{4.23}$$

To obtain the equation of the trajectories, we have to integrate:

$$\varphi - \varphi_0 = \int_{r_0}^{r} \frac{\Delta dr}{r \sqrt{n^2 r^2 - \Delta^2}}. \tag{4.24}$$

For an index of refraction of the form (4.23), the equation for the light rays,

$$r^2 - r \frac{\sqrt{1 - \Delta^2}}{\Delta} \cos \varphi - 1 = 0,$$

are closed curves with centers of inversion shown in Fig (2.17).

The closed curves on the sphere intersect at antipodal points. All light rays that emanate from one of the points will coalesce at the other, inverse point. In this way Maxwell constructed a perfect, undistorted image of the other. But, he did more.

From an analysis of motion on the sphere, Maxwell (1854) could determine both the dynamics (motion in a Coulomb potential), and geometry (either circular or elliptic orbits) in the plane.

It is possible to go beyond Maxwell. Based on the same conformal analysis as that of Maxwell' s fish-eye, Luneburg [14] was able to generalize Maxwell's fish-eye index of

---

[14] op. cit. p 178ff.

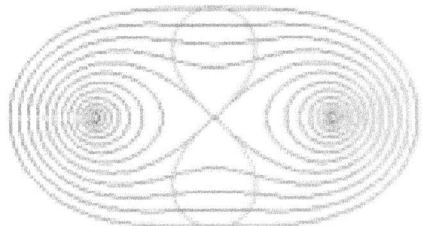

Figure 4.4: Cassini's ovals consist of a set of points in the plane for which the product of the
distances to two fixed points is constant.

refraction to:

$$n(r) = \frac{2\gamma r^{\gamma-1}}{1 + r^{2\gamma}} \quad \gamma \geq 1, \tag{4.25}$$

which reduces to it for $\gamma = 1$. Introducing (4.25) in the next generalization of $\gamma = 2$
into (4.24), gives the equation of the trajectory as:

$$r^4 - 2a^2 r^2 \cos(2\varphi) + a^4 = b^4. \tag{4.26}$$

This constitutes the next step after ellipses. Whereas the *arithmetic* mean of the distances from two focal points, r and r', to a point on the ellipse remains constant, the *geometric* mean of the distances, $b = rr'$ remain constant for Cassini ovals, as seen in
Fig (4.4).

In other words, Cassini ovals are loci of points for which the geometric mean, b, of the
distances from the foci to any point on the oval is constant.

Instead of cutting a cone by a plane, one slices a torus with a plane, shown in Fig (4.5).
As can be seen from Fig (4.4), Cassini ovals have four characteristics forms that depend
on the ratio of the coefficients a and b in Eq (4.26). When $b \geq a$, the Cassini oval is
a convex curve similar to a Keplerian ellipse. When $a < b < 2a$, a concave bridge
appears, and when $a = b$ the bridge closes and the Cassini curve transforms into
an inverted '8' that Jacob Bernoulli rediscovered about a fourteen years later (1694),
which is named after him, *lemniscate* (Latin for pendant ribbon) of Bernoulli. It is
defined as the locus of points the product of whose distances from two fixed points, or
foci, separated by a distance $2a$, is constant, a^2.

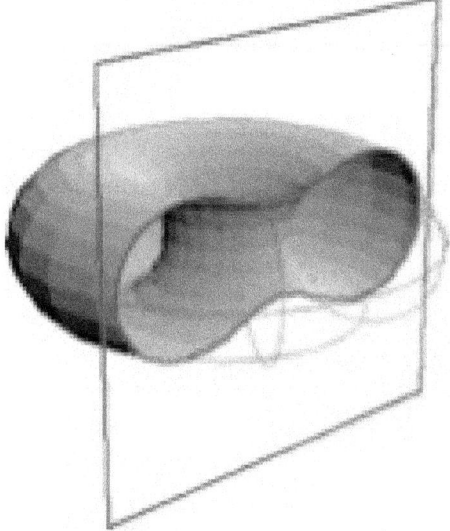

Figure 4.5: Cassini ovals as planar sections of a torus. Spiric (torus=ancient Greek) sections
are obtained by slicing a torus by a plane.

Figure 4.6: Three equal masses orbiting in a lemniscate at uniform distance from one another
in a stable orbit.

Three-bodies rotating so as to remain constantly in line was first discovered by Euler. However, such a three-body system is unstable. However, two bodies moving closely around each other while both of them move around a third body further away mimics the Earth-Moon-Sun, and is stable. Such a configuration obeys Hill's equation. Recently, a fourth solution has been found where three bodies of equal mass follow each other, keeping a uniform spacing, in a lemniscate orbit, as shown in Fig (4.6). Such stable solutions are referred to as 'choreographies,' which were first predicted by Moeckel [15] in 1988, and confirmed through computer calculations by Moore [16], five years later.

Finally, when $a > b$, the Cassini oval splits into two eggs that are symmetrical with respect to the line of symmetry.

The orthogonal trajectories of Cassini's ovals are rectangular hyperbolas. They are equipotential lines of the electrostatic field that have been created by two uniformly charged parallel wires consisting of identical charges. Or, they can be the field lines of the magnetic field that are formed by two parallel wires carrying equal intensity currents in the same direction. In any plane orthogonal to the wires Cassini ovals can be seen with foci where they intersect with the wires, as shown in Fig (4.7).

---

[15]R. Moeckel, "Some qualitative features of the three-body system," in *Hamiltonian Dynamical Systems.* K R Mayer &
D G Saari, eds (AMS, Providence RI, 1988).
[16]C Moore, "Braids in classical gravity," *Phys Rev Lett* **70** (1993) 3657-3679.

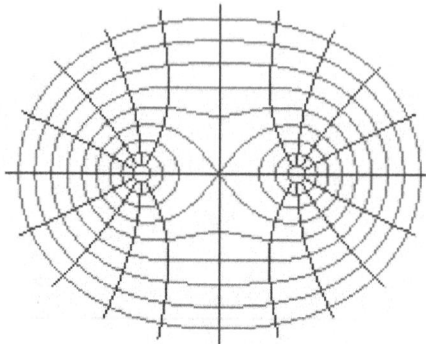

Figure 4.7: The field lines of the magnetic field that are due to two parallel wires transporting currents in the same direction when viewed in any plane orthogonal to the wires appear as Cassini ovals with foci which intersect with the wires.

Also the field lines of the magnetic field created by two parallel wires carrying currents of equal intensity in the same direction are Cassini ovals with foci at the intersection points of the wires. Electro-static and -dynamic configurations can be found in the plates to volume II of Maxwell's *A Treatise on Electricity and Magnetism*.

One may wonder why Cassini, who was an accomplished astronomer in his own right, would want to replace Keplerian ellipses, by a more unsymmetrical close curve. It was known that the planets describe elliptical orbits of low eccentricity, the angular sectors of rays traced out at one focus would be almost the same as that at the other focus.

Cassini imagined that the Sun travelled around the earth on one of these ovals, with the earth at the focus of the oval, i.e. the antithesis of Newton. The Sun revolves around the earth in such a way that if you draw two lines from the center of the sun to the two foci, the rectangle defined by these lines is always equal to the rectangle defined by the larger and smaller distances between the Sun and the Earth.

Cassini studied his curves in 1680; over a century later, d'Alembert, writing in his *l'Encyclopedie* article concluded

This curve, that Mr Cassini had wished to introduce into astronomy, is now only a purely geometric curve, and an object of curiosity, because we know that planets describe Apollonian or ordinary ellipses.

More recently, such spiric (torus) curves have been implicated in a, no less, fourth rank tensor theory of gravity that would fix many of the woes of GR. [17] Such is the bug that Einstein has spread.

So whether we are talking about conic sections or cassinoids, both can be considered as a generalization of the polar equation:

$$\varphi = \int \frac{\Delta dr}{r\sqrt{r^2 - \Delta^2}} = \sin^{-1}(\Delta/r), \qquad (4.27)$$

of a straight line. An analogous integral can be written in terms of the 'turning' parameter,

$$\Delta \int \frac{d\Omega}{\sqrt{\rho^2 - \Delta^2}} = \Psi(\Delta),$$

where $\Omega = \ln \rho$. To find $\Psi(\Delta)$ we employ an Abel transform,

$$\Omega(\rho) = \frac{2}{\pi} \int \frac{\Psi(\Delta)d\Delta}{\sqrt{\Delta^2 - \rho^2}}. \qquad (4.28)$$

Strangely enough, the two integrals are related by the expression for the total angle of deflection:

$$\chi = \pi - 2\phi_\infty, \qquad (4.29)$$

shown in Fig (3.1), where ϕ_∞ is the angle that the ray makes at its closest point to the target.

Since $\phi_\infty = \cos^{-1}(\Delta)$ when the limits on the integral (4.28) are taken from Δ to 1, the trigonometric expression,

$$\cos^{-1}(\Delta) = \pi/2 - \sin^{-1}(\Delta),$$

---

[17] J A Nieto & L A Beltran, *Toward an alternative gravitational theory* arXiv:1406.0779.

gives

$$\chi/2 = \sin^{-1}(\Delta).$$ (4.30)

We can express this as the following equivalence,

$$\frac{2}{\pi}\int_{\Delta}^1 \frac{\cos^{-1}(\Delta)d\Delta}{\sqrt{\Delta^2 - \rho^2}} = \int_{\Delta}^1 \frac{d\Delta}{\sqrt{\Delta^2 - \rho^2}} - \frac{2}{\pi}\int_{\Delta}^1 \frac{\sin^{-1}(\Delta)d\Delta}{\sqrt{\Delta^2 - \rho^2}} = -\ln \rho.$$

Since the first integral on the right-hand side is

$$\int_{\Delta}^1 \frac{d\Delta}{\sqrt{\Delta^2 - \rho^2}} = \ln\left(\frac{1 + \sqrt{1 - \rho^2}}{\rho}\right),$$

we obtain the second integral as:

$$\frac{2}{\pi}\int_{\Delta}^1 \frac{\sin^{-1}(\Delta)d\Delta}{\sqrt{\Delta^2 - \rho^2}} = \ln(1 + \sqrt{1 - \rho^2}).$$

If we replace the arcsine by the total angle of deflection, $\chi = \pi/\gamma$, with $\gamma > 0$, then

$$1 + \sqrt{1 - \rho^2} = \left(\rho^{-1} + \sqrt{\rho^{-2} - 1}\right)^{1/\gamma}.$$ (4.31)

These expressions can be equated with r.

The beautiful conclusion we arrive at is that when r is set equal to the left-hand side
of (4.31) we get:

$$n^2 = \frac{2}{r} - 1,$$ (4.32)

our old friend, the Eaton index of refraction, (2.48) for $a = 1$. And when it is set equal
to the right-hand side, there results:

$$n = \frac{2r^{\gamma-1}}{1 + r^{2\gamma}},$$

which is Maxwell's fish-eye, and all it generalization including Cassini ovals for $\gamma = 2$.
Indeed, Newton's law of attraction reigns supreme!

As we mentioned, Newtonian index of refraction, (4.32), corresponds to the Eaton lens at an angle of refraction of π radians. Its *negative* is the inverse of index of refraction of the Schwarzschild metric [18]

$$n^2 = (1 - \alpha/r)^{-1},$$

in 'natural' units where $c = 1$. This is not to say that materials with *negative* indices of refraction don't exist but they are a special brand, known as metamaterials, that flat lenses with a negative index of refraction make perfect lenses. [19] Such a medium is called a left-handed medium, and requires both negative permettivities and permeabilities. But, this is not what Moller had in mind.

Another point is the sign of χ. Negative χ's bend outward, not inward as in Moller's exposition of the deflection of light. [20]

4.6 Deflection of light on a pseudosphere

The equation of light rays in a medium of index of refraction,

$$n = \frac{1}{1 + r^2},$$

gives rise to an equation for the light rays as:

$$r^2 - r \frac{\sqrt{1 - \Delta^2}}{\Delta} \cos \phi - 1 = 0.$$

The product of any two solutions to this equation is

$$rr' = -1.$$

From this, Maxwell concluded that to any sphere of radius, r, there is associated a conjugate sphere of radius $r' = 1/r$ which is a perfect, undistorted optical image of the sphere of radius r'. The image is inverted and its magnification is given by the

---

[18] Moller, *op. cit.*, p 355, Eq (54).

[19] J B Pendry, "Negative refraction makes a perfect lens," *Phys Rev Lett* **85**, 3966.

[20] *op. cit.*, p 354, Eq (52). Light, like matter, should be attracted to mass, not repulsed!

Figure 4.8: The conformal metric (4.33) can be interpreted as the line element of a sphere with an index of refraction, (4.34). Points of a unit sphere are mapped, by stereographic projection, into circles in the plane. Taken from youtube fsgm1 (17/12/2010) "Stereographic projection of a Riemann sphere."

ratio of the radii. *Light waves in this medium are stereographic images of geodesic curve on a sphere, i. e. great circles. Light can travel in circles, but the gravitational field which caused them to do so cannot.*

The conformal line element:

$$ds^2 = n^2(r)(dx^2 + dy^2),$$ (4.33)

will be that for a line element of a sphere for an index of refraction:

$$n(r) = \frac{2}{1 + r^2}.$$ (4.34)

The medium in the plane will, however, have an index of refraction:

$$n^2 = \frac{1}{r} + C$$ (4.35)

which represents a Coulomb field with negative total energy, $-C$.

For the deflection of light by massive bodies, elliptic orbits become hyperbolic so that the same stereographic projection will not apply. Since the sphere has constant, positive curvature we look for its analog with constant, negative curvature.

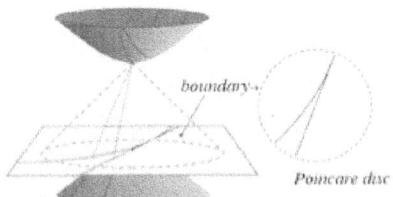

Figure 4.9: A pseudosphere is used to describe a surface of constant negative curvature. However, it is "incomplete" insofar as it cannot be embedded into $3D$ space. It's projection on the plane gives the Poincaré disc model.

The line element should be conformal, and therefore be given by (4.33) except with an index of refraction

$$n^2 = \frac{2}{1 \pm r^2}. \tag{4.36}$$

This line element was written down by Riemann in 1854; with the plus sign it corresponds to a stereographic projection from a sphere onto a plane, as shown in Fig (4.8). Alternatively, for the negative sign it corresponds to stereographic projection from a 'pseudo' sphere, which was only discovered later by Beltrami in 1865. Poincaré used (4.36) as a model of hyperbolic geometry with constant curvature, one that could be derived from his half-plane model. So if it is stereographic, so, too, was is its projection and what is the index of refraction in the plane?

The stereographic projection from the pseudosphere is shown in Fig (4.9). The geodesics in the plane are no longer straight lines, but, curved arcs which cut the boundary of the Poincarè disc orthogonally.

The index of refraction in the medium is:

$$n^2(r) = C - \frac{2\alpha}{r}, \tag{4.37}$$

with the total energy, $C > 0$ and α is the Schwarzschild radius. It is the negative of (4.36), and the inverse of Moller's index of refraction implicating the coefficient in the Schwarzschild metric.

We have here a classical scattering problem with an impact parameter, $\Delta = L/C$.

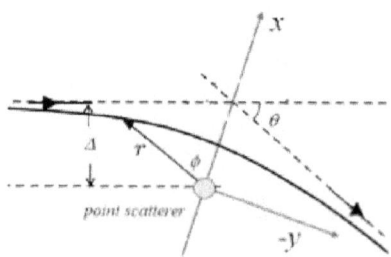

Figure 4.10: The classical scattering configuration for a point scatterer.

From (4.24) we can immediately write down equation of the trajectory as:

$$\phi - \pi = \int_{r_0}^{r} \frac{\Delta dr}{r\sqrt{r^2 - 2\alpha r - \Delta^2}}.$$

The solution is a hyperbolia,

$$r = \frac{\Delta}{1 + \varepsilon \cos \phi},$$

with eccentricity

$$\varepsilon = \sqrt{1 + (\Delta/\alpha)^2}.$$

The scattering angle, as shown in Fig (4.10), is easily found to be

$$\cot(\phi/2) = \sqrt{\varepsilon^2 - 1},$$

which for small angles, reduces to

$$\phi = 2\alpha/\Delta. \qquad (4.38)$$

This is formally what GR finds, but with the important exception that the impact parameter is replaced by the radius of the sun.

Such an equivalence would lead to an angular momentum per unit mass of no less than $10^{17} m^2/s$! One may ask, angular momentum of what? Without the radius of the sun, the right deflection angle would not come out. The factor of 2 in (4.38) is a red-herring. Without it you would get Einstein' s pre-GR result; with it you get what has been heralded as a major triumph of GR over classical theory. However, without the gravitational potential you have no scattering at all, so the GR result cannot be considered as a 'perturbation', perturbation of what?

4.7 Combining static & motional forces

In the chapter on GO and GR we have appreciated that the same formulas can be given entirely different meanings, and the same physical phenomena can be explained in many different ways. Two such examples are the deflection of light by a massive body and the advance of the perihelion of Mercury. Both these phenomena were known before Einstein wrote down his field equations, and even the gap in the magnitude of the perihelion was known, but not to the degree which it is known today. [21]

In electromagnetism, there is a dynamical force which is the synthesis of Coulomb's static force and Ampère's angle law,

$$\frac{ids \cdot i'ds'}{r^n}\Phi(\vartheta, \vartheta'),$$

where ds and ds' are two current elements with intensities i and i', respectively, and ϑ and ϑ' are the angles they form with the line connecting them. The determination of the exponent, n, required an equilibrium experiment on the part of Ampère. However, if the force is to reduce to Coulomb's in the static case then $n = 2$.

Weber's force replaces the current elements by the product of the velocity, v and charge, e, $ve = cids$, where the constant of proportionality, c, i "number of units of statical electricity which are transmitted to the unit electric current in the unit of time. [22]

This value differed from Weber's constant c', which, according to him, is "that relative velocity which electrical masses e and e' have and must retain if they are not to act on each other at all."

[21]A A Vankov, "General relativity problem of Mercury's perihelion advance revisited,"arXiv:1008.1811.
[22]J Clerk Maxwell, *A Treatise of Electricity & Magnetism*, (Clarendon Press, Oxford, 1873) vol II, §1849.

His fundamental law of force was (2.14). In 1856, Weber and Kohlrausch performed experiments to determine the value of c', which they found exceeded Maxwell's constant, c, by the amount $\sqrt{2}$.

How this arises can be seen by considering the Lagrangian,

$$\mathcal{L} = -r(c\dot{t})^2 + \frac{\dot{r}^2 + r^2\dot{\varphi}^2}{r},$$

in the plane $\vartheta = \pi/2$. The Euler-Lagrange equations,

$$\frac{d}{d\tau}\frac{\partial\mathcal{L}}{\partial\dot{x}} - \frac{\partial\mathcal{L}}{\partial x} = 0,$$

are explicitly given by

$$2\frac{d}{d\tau}\left(\frac{\dot{r}}{r}\right) + \frac{\dot{r}^2}{r^2} - \dot{\varphi}^2 + (c\dot{t})^2 = 0,$$

$$\frac{d}{d\tau}r\dot{\varphi} = 0 \qquad \text{and} \qquad \frac{d}{d\tau}rc^2\dot{t} = 0.$$

The last two equations lead immediately to the first integrals, $r\dot{\varphi} = v = \text{const.}$,

$$r\dot{t} = H = \text{const}, \tag{4.39}$$

of the motion. Whereas the former has the significance of a velocity, the latter does not so we can set H equal to unity. In other words, $\dot{t}$, has no physical significance, and its only role is to introduce a limiting velocity, c.

Upon introducing the first integrals into the radial equation, there results:

$$2\frac{\ddot{r}}{r} - \frac{\dot{r}^2}{r^2} + \frac{c^2 - v^2}{r} = 0.$$

This is precisely the condition that Weber's force vanish, which was to be expected since it resulted from the free Euler-Lagrange equations. It differs from Weber's formulation in that $\sqrt{c^2 - v^2}$ sits in for Weber's c'. Notwithstanding the fact that Weber's c' is less than c, it points to the fact that c is a limiting velocity, and its decrease is due to the contraction of the charges in the direction of motion. This pre-dates Lorentz by almost half a century. [23]

We now compare this with the determination of the angle of deflection when light. The relevant lagrangian is:

$$\mathcal{L} = \frac{1}{2} \left(\frac{\dot{r}^2}{(1 - \alpha/r)^2} + \frac{r^2 \dot{\varphi}^2}{1 - \alpha/r} - c^2 \dot{t}^2 \right),$$

whose space component is the Beltrami metric. This is obtained by dividing the Schwarzschild metric,

$$-c^2 d\tau^2 = -(1 - \alpha/r)c^2 dt^2 + \frac{dr^2}{1 - \alpha/r} + r^2 d\varphi^2,$$

through by $(1 - \alpha/r)$ and $d\tau^2$. The Lagrangian would then be half of the right-hand side. To sum up, the Schwarzschild metric is conformally equivalent to the Beltrami metric.

Subtraction of a constant from a Lagrangian would have no effect on the Euler-Lagrange equations. But, if we were to consider the resulting Lagrangian by merely dividing through by $d\tau^2$,

$$-c^2 = -(1 - \alpha/r)c^2 \dot{t}^2 + \frac{\dot{r}^2}{1 - \alpha/r} + r^2 \dot{\varphi}^2,$$

and considering the Lagrangian to be half the right-hand side, we would obtain not only the conservation of momentum, but, in addition, we get:

$$(1 - \alpha/r)\dot{t} = c/v = \text{const.}$$

as a first integral of the motion. Unlike mechanics where static potentials appear in the space part of the metric, these potentials now appear in the time component because forces don't exist in GR.

The introduction of a static potential into the force law goes hand-in-hand with the introduction of a limiting velocity. Although the appearance of a limiting velocity gives some credence of treating two times, appearances can be deceiving!

---

[23] This is not the first time Lorentz did not give recognition where recognition was due. His transforms were derived by Voigt in 1887, but, in this case he gave a late acknowledgment in *The Theory of Electrons*.

Introducing the first-integrals into the radial equation, dividing through by $d\varphi^2$, and changing the integration variable to $u = 1/r$, lead to the equation of the trajectory as:

$$\varphi - \varphi_0 = \int_{u_0}^{u} \frac{du}{\sqrt{W + \alpha u/L^2 - u^2 + \alpha u^2}}, \qquad (4.40)$$

where

$$W = \frac{(c/v)^2 - 1}{L^2}.$$

Closed orbits require a negative total energy, $W < 0$, which seemingly violates the fact that c is a limiting velocity. Moreover, the linear term, αu, wouldn't be there if we didn't consider proper time. Why should time considerations be necessary to introduce a static potential?

Conventional wisdom claims [24]

> Unfortunately, even though this [Eq (4.40)] is an exact solution to the problem, it expresses the angle φ as an integral of $u = 1/r$... To make the problem more transparent, differentiate with respect to u to make it look like a classical Kepler problem.

Why bother? The use of the relation, $\dot{r}/\dot{\varphi} = dr/d\varphi$ contradicts the fundamental assumption that "the field of a static body cannot depend on time which is ensured by $R_{01} = -\dot{A}/rA$, which implies $A = A(r)$." Here, R_{01} is the time component of the Ricci tensor and $A = 1 - \alpha/r$.

The metric coefficients have to be functions of t if r is. Moreover, the need for introducing a static potential through a time coefficient is absurd. The only reasoning for introducing time is when velocity, and acceleration components are appended to an otherwise static potential, as Weber has taught us. Time appears only in the derivatives of the converted Ampère law. There is no rhyme or reason to the methodology of GR.

---

[24]Adler *et. al., op. cit.*

We may reason from electrodynamics. The introduction of Coulomb's law into the kinetic formulation of Ampère's law does not simply include that potential in the coefficient of the time component of the metric. Rather, what appears in the time component is a constraint stipulating that c is the limiting velocity.

Helmholtz launched a scathing attack on the Weber force by claiming that two charged particles which have finite velocities, and are a finite distance from one another may perform an infinite amount of work through an infinite kinetic energy. Here, a limiting velocity is to no avail.

Helmholtz considers a fixed non-conducting spherical surface with radius a and has a surface density σ (C/m^2). A particle of mass m and charge e is moving with velocity v. From the Weber potential,

$$V_W = \frac{ee'}{r} \left[1 - \frac{\dot{r}^2}{c'^2} \right],$$
(4.41)

Helmholtz determines the electrodynamic potential as:

$$4\pi a \sigma e \left(1 - \frac{v^2}{3c'^2} \right)$$
(4.42)

to which he adds the kinetic energy, $mv^2/2$ to get the equation of energy conservation:

$$\frac{m}{2} \left(1 - \frac{4\pi}{3} \frac{ae}{mc^2} \right) v^2 + 4\pi a \sigma e = \text{const.}$$
(4.43)

Maxwell [25] makes a valent attempt to refute Helmholtz's criticism by which he argues that the second term in the coefficient of v^2 may be increased indefinitely by increasing a so as to render the coefficient negative. However, Maxwell could not have possibly known that the limiting speed of light also imposes fundamental lengths, whether it be in electromagnetism or gravitation.

In Maxwell's words:

---

[25] Maxwell, *op. cit.* §856 p 431.

Acceleration of the motion of the particle would then correspond to a diminution of its *vis viva*, and a body moving in a closed path acted on by a force like friction, always in opposite direction to its motion, would continually increase in velocity, and that without limit. This impossible result is a necessary consequence of assuming any formula for the potential which introduces negative terms into the coefficient of v^2.

What Maxwell failed to realize that the second term in the coefficient of v^2 is the ratio of the classical electron radius to characteristic length of the non-conductor, $1/(4\pi/3)a\sigma$.

If we replace the surface charge density, σ, by surface gravity, g, the above line of reasoning would imply that for planets whose radii are greater than

$$R = c^2/g, \tag{4.44}$$

would be torn apart because the surface gravity would be too small to hold it together.

The coefficient of v^2 in the energy conservation equation (4.43) is analogous to the inverse coefficient of dr^2 in the Schwarzschild inner solution, where R in (4.44) is the radius of the disc. Nothing exists outside the disc because exceeding the rim, which would take an infinite amount of time, the planet would be torn apart.

A similar criterion can be obtained for the outer solution. Since gravity is attractive, the Riemann potential: [26]

$$V_R = \frac{\mu m}{r}\left(1 + \frac{\dot{r}^2}{c^2}\right) \tag{4.45}$$

The gravitational potential calculated from (4.45)is:

$$\frac{\mu m}{a}\left(1 + \frac{v^2}{c^2}\right),$$

where m is the peripheral mass. Now, *subtracting* this from the kinetic energy of the peripheral mass, mv^2, gives the conservation equation:

$$\frac{1}{2}\left(1 - \frac{2\mu}{ac^2}\right)v^2 - \frac{\mu m}{c} = \text{const.}$$

[26]A O'Railly, *Electromagnetic Theory*, vol II (Dover, New York, 1965) p 527 Eq (11.11).

It is clear that the radius cannot be less than

$$R = 2\mu/c^2, \tag{4.46}$$

for, otherwise, the coefficient of the kinetic energy would be negative. The same deductions apply as above, namely, a system with negative kinetic energy would continually increase in velocity when acted upon forces, like friction, that normally lead to a decrease in velocity.

Hence, the Schwarzschild radius (4.46) is the smallest possible radius. Again the coefficient of the kinetic energy appears in the inverse coefficient of the dr^2 term in the metric, and (4.46) represents the boundary which cannot be superseded.

As we saw in Chapter 1, the speed of light not only introduces a limit for the maximum speed, but it, moreover, introduces a fundamental length.

In electrodynamics it is the classical electron radius that represents a lower bound, while in gravitation, it is the Schwarzschild radius that presents the same lower bound to the radius. These are bounds that no classical theory can supersede. All bounds on the size of the radial coordinate depend on a limiting velocity. Surface gravity like the gravitational potential are both limited by a limiting speed of propagation.

Weber's force can be derived from the metric

$$-c^2 d\tau^2 = -rc^2 dt^2 + \frac{dr^2}{2r^2}. \tag{4.47}$$

The corresponding Lagrangian is:

$$\mathcal{L} = \frac{1}{2} g_{\nu\mu} \dot{x}^\nu \dot{x}^\mu = \frac{1}{2} \left(-rc^2 \dot{t}^2 + \frac{\dot{r}^2}{r} \right).$$

Coulomb's potential is buried in the first term.

Weber's force will be given by the Euler-Lagrange equation:

$$F_W/ee' = \frac{d}{d\tau} \frac{\partial \mathcal{L}}{\partial \dot{r}} - \frac{\partial \mathcal{L}}{\partial r},$$

which is explicitly given by:

$$F_W = ee' \left(\frac{\ddot{r}}{r} - \frac{\dot{r}^2}{2r^2} + c^2 \dot{t}^2 \right).$$

Time t is the only cyclic variable since we are considering radial motion. The first integral is given by (4.39), and introducing it into the radial equation gives Weber's law,

$$F_W = \frac{ee'}{r^2} \left(r\ddot{r} - \frac{1}{2}\dot{r}^2 + c^2 H^2 \right).$$

What then is the significance of the coefficient of dt^2 in the metric (4.47)?

It is well known [27] that Weber's force can be derived from the Lagrangian,

$$\mathcal{L} = \frac{ee'}{r} \left(1 + \frac{\dot{r}^2}{c^2} \right),$$

where the limiting velocity is introduced explicitly into the space component of the metric. So why go through the bother of introducing a time component into the Lagrangian?

Although the limiting velocity c appears more transparent in the time component, it will not be the Coulomb potential which appears in that expression. There is a constraint that must be satisfied. The potential that will appear in the time component represents the surface charge, σ on a ball of radius r whose surface area is S:

$$\Phi = \sigma \cdot S = \frac{1}{4\pi\epsilon_0 r} \cdot 4\pi r^2 = \frac{r}{\epsilon_0}.$$

4.8 Unifying electromagnetics & gravitation

Einstein's GR is a field theory of gravitation, and, in a (unsuccessful) attempt of unification with electromagnetism, he naturally turned to Maxwell's field theory. The only

---

[27] A O'Rahilly, *op. cit.* p 526.

place where a material force appears in the field equations of electromagnetism is in the Lorentz force which is used to bridge the gap between particle dynamics and the electromagnetic field.

Weber's electrodynamics, based upon his force law describing the interaction of two infinitesimal current elements, fell out of favor after Hertz's discovery of electromagnetic waves in 1884. That discovery ushered in Maxwell's field theory on the continent which gained almost universal acceptance.

In many respects, the two theories of electrodynamics are complementary. Weber' s force is a longitudinal force acting along the radial vector connecting the two current elements. Maxwell's theory has fields normal to the direction of motion, and normal to themselves in free space.

Lorentz's podermotive force was discovered a half-century before by Grassmann (1845) who envisioned a transverse force that fails to satisfy Newton's third law of action-and-reaction. This was, of course, passed on to Lorentz's force to which the likes of J J Thomson reprehensible.

Yet, it was more amenable to the nascent theory of special relativity which was supported by Kaufmann's experiments on the dependence of an electron's mass upon its speed. The so-called transverse mass fit his experiments much better than the longitudinal mass which has long been deceased.

It was not long after the ink dried on Weber's paper in which he derived his law of force, that astronomers on the continent began thinking of how to apply it to gravitation. The attempt was to place gravitation under the banner of electrodynamics.

It all began with a hypothesis of Mossotti who tried to account for gravitational attraction by assuming that attraction between opposite charges was slightly greater than the repulsion between like charges. Weber's law would not only account for Newtonian attraction, but, would moreover describe the relative velocities and relative accelerations of pairs of heavenly bodies.

The advances of the perihelion of Mercury and Venus were already known from the observations of Le Verrier and others. Then came a slew of attempts to apply Weber's law by Scheibner who determined the secular variation of the perihelion of Mercury to be 6.73 arcsec, while Tisserand found 6.28 arc-seconds, and 1.32 arcsec for Venus. Has anyone ever asked the question of why GR insists on Mercury alone? [28]

Even as late as 1925, people were still trying to fix the parameter, γ in

$$\Delta = 2\pi\gamma \left(\frac{\mu}{cL}\right)^2, \tag{4.48}$$

to account for the advance in the perihelion of Mercury. Even Erwin Schroedinger [29] got into the fray, but not for the reason that Barbour claims, The "main value of the paper [is] its potential to serve as a simple illustration of GR." Rather, it appears that Schroedinger, using the Weber potential, (4.41), had every intention of sidetracking GR, show that the advance of the perihelion of Mercury would be obtained with resorting to it.

Yet, there is a peculiarity of Weber's law based on what Ampère attempted to describe. In his own words:

> An infinitely small portion of an electrical current exerts no action on another infinitely small portion of current situated in a plane which passes through its midpoint, and which is perpendicular in direction.

Ampère expressed his force law between two current elements, ids and $i'ds'$ in general terms as:

$$\frac{ids \cdot i'ds'}{r^n}\Phi(\vartheta, \vartheta'),$$

where Φ was an unknown function of the angles ϑ and ϑ' are the angles that the current elements make with the line connecting them. The unknowns were n, the exponent, and Φ which he had to determine through experiment.

---

[28]C Poor, "Relativity and the motion of Mercury," *Annals N Y Acad Sci* **XXIX** (1925) 285-319.

[29]E Schroedinger, "The possibility of fulfillment of the relativity requirement in classical mechanics," *Ann d Phys* 77(1925) 325-336; translated in *Mach's Principle: From Newton's Bucket to Quantum Gravity* J B Barbour & H Pfister eds (Birhauser, Boston, 1995), p 147.

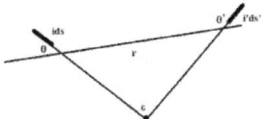

Figure 4.11: Geometrical configuration of current elements lying in the same plane. They
make angles ϑ and ϑ' with respect to the line connecting them, and an angle ε
with respect to each other.

He then separated the force components into transverse, $\sin\vartheta ds$ and $\sin\vartheta' ds'$, and
longitudinal, $\cos\vartheta ds$ and $\cos\vartheta' ds'$, components. With ω as the angle of inclination
between one plane containing one of the current elements and the other, Ampère ex-
pressed the angular dependencies in Φ as:

$$d^2 F_A/dsds' = \frac{ii'}{r^n}(\sin\vartheta \sin\vartheta' \cos\omega + k\cos\vartheta \cos\vartheta'),$$

reducing the number of unknowns to n and k.

For angles of tilt $\omega \neq 0$, his force law implies spherical, rather than Euclidean geometry.
This was rather superfluous since the components can always be separate into parallel
and perpendicular components. The latter can be excluded because there is no force
between perpendicular components according to Biot-Savart. The configuration of the
current elements is shown in Fig (4.11).

The parameter, k, was taken by Ampère to be the ratio of longitudinal to transverse
forces. Ampère's equilibrium experiments found it to be $k = -1/2$, and n he set equal
to 2, undoubtedly based on the universality of the inverse square laws like Coulomb
and Newton.

The choice $k = -1/2$ reduces the component of the force parallel to one circuit element to a total derivative one of the circuit elements so that when it taken around a
loop, it will vanish. The components in the other two, orthogonal, directions coincide
with the law of Biot-Savart, which preceded Ampère work.

Ampère's force law could thus be expressed as:

$$d^2 F_A / ds ds' = \frac{ii'}{r^2} \left(\cos \varepsilon - \frac{3}{2} \cos \vartheta \cos \vartheta' \right),$$

where the difference in angles, $\varepsilon = \vartheta' - \vartheta$, satisfies

$$\cos \varepsilon = \cos \vartheta \cos \vartheta' + \sin \vartheta \sin \vartheta' = \frac{\partial}{\partial s'} \left(r \frac{\partial r}{\partial s} \right) = \frac{\partial r}{\partial s} \frac{\partial r}{\partial s'} + r \frac{\partial^2 r}{\partial s \partial s'}.$$

The relations, $\cos \vartheta = \partial r / \partial s$ and $\cos \vartheta' = \partial r / \partial s'$ bring to mind the dot product, while $L = r \sin \vartheta$ and $L' = r \sin \vartheta'$ are reminiscent of the vector product representing angular momenta per unit mass where $\sin \vartheta = r d\vartheta' / ds$ and $\sin \vartheta' = r d\vartheta / ds'$.

With these substitutions, Ampère's law takes on the suggestive form:

$$d^2 F / ds ds' = \frac{ii'}{r^2} \left(\frac{1}{2} \frac{\partial r}{\partial s} \frac{\partial r}{\partial s'} - \frac{L \cdot L'}{r^2} \right),$$

as an inverse square law whose coefficient is the difference between non-relativistic kinetic energy and angular momenta. In contrast,

$$\cos \varepsilon = \frac{\partial r}{\partial s} \frac{\partial r}{\partial s'} + \frac{L \cdot L'}{r^2},$$

represents the sum of twice the kinetic energies and angular momenta. Does the fact that the non-relativistic kinetic energy appears in Ampère's law (which is a consequence of $k = -1/2$) limit the fundamental validity of the law?

That is to say that gravity does not travel in closed loops like electric currents, although a 'gravito-magnetic' force appears necessary to generate transverse gravitational waves that travel at the same speed as light. We will have much more to say about this soon, but consider, for the moment, two examples.

If the current elements both lie along the radial vector connecting the, $\vec{r}$, all angles are zero, and as well as the angular momenta. Ampère's force law then is negative. Alternatively, if the two current elements are parallel to one another, $\varepsilon = 0$ and $\vartheta = \vartheta' = \pi/2$. Now, the force is attractive, and twice in magnitude as in the previous

example. Consequently, two parallel elements carrying the same current in the same
direction attract each other with twice the force that two elements lying along the
same line repel each other. Gravity knows no such thing so it is not a matter and
exchanging charges for mass and permittivity for the gravitational constant.

In fact, Ampère's law can be cast in the form of a non-relativistic virial theorem. Consider the following relation:

$$\frac{1}{2}\frac{\partial^2 r}{\partial s \partial s'} = \frac{\partial}{\partial s'}\left(r\frac{\partial r}{\partial s}\right) = r\frac{\partial^2 r}{\partial s \partial s'} + \frac{\partial r}{\partial s}\frac{\partial r}{\partial s'} = \cos\varepsilon.$$

As Maxwell shows, the cosine of the angle between the circuit elements is related to
magnetic induction . Now, subtracting 3 times the kinetic energy,

$$\mathcal{T} = \frac{1}{2}\frac{\partial r}{\partial s}\frac{\partial r}{\partial s'},$$

from both sides of the equation results in:

$$\frac{1}{2}\frac{d^2\mathcal{I}}{ds ds'} = 3\mathcal{T} + \frac{1}{ii'}\frac{d^2 F_A}{ds ds'}r^2, \tag{4.49}$$

where $\mathcal{I}$ represents the moment of inertia per unit mass .

Equation (4.49) is Ampère's law written as a virial. In comparison, the mechanical,
non-relativistic virial is:

$$\frac{1}{2}\frac{d^2\mathcal{I}}{ds ds'} = 2\mathcal{T} + Fr, \tag{4.50}$$

where F is a mechanical force per unit mass that satisfies:

$$F = \frac{\partial^2 r}{\partial s \partial s'}. \tag{4.51}$$

The difference in numerical factors multiplying the kinetic energy leads us to believe
that there are possibilities other than Ampère's choice of $k = -1/2$.

However, Ampère's law plays a prominent role in electromagnetism since it leads to
the conclusion that the force exerted by a complete circuit on a current element is
perpendicular to that element.

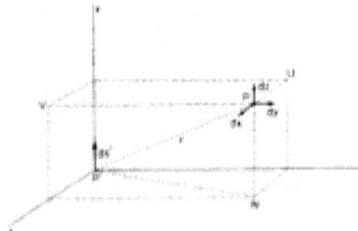

Figure 4.12: The line element $i'ds'$ situated at the origin P' acts on another line element, ids, at P. The element which does the acting is supposed to be directed along the z-axis. Taken from R A R Tricker, *Early Electrodynamics: The first law of Circulation* (Pergamon, Oxford, 1965) Fig 12.

Consider the configuration shown in Fig (4.12).

The angle between the two elements, ε is given by $\cos\varepsilon = dz/ds$, and the angles that the element make with r is $\cos\vartheta = dr/ds$, and $\cos\vartheta' = z/r$. [30] The component of the force in the z-direction is:

$$\frac{z}{r}d^2F/dsds' = \frac{ii'z}{r^3}\left(\frac{dz}{ds} - \frac{3}{2}\frac{dr}{ds}\frac{z}{r}\right) = \frac{1}{2}ii'\frac{d}{ds}\left(\frac{z^2}{r^3}\right).$$

When this total differential is integrated around a closed circuit it vanishes, thereby demonstrating the principle that forces exerted by a closed circuit a current element are at right angles to that element. The invariant is $\cos^2\vartheta'/r$. The components of the force in the x- and y- directions give the law of Biot-Savart.

Current elements don't exist in isolation, and neither do masses: you need another mass to get gravitational energy. But, gravity is unlike charges that are in motion that create a force when the circuit is closed. Therefore, Ampère's result that the ratio of the force between parallel current elements is half that between longitudinal elements may be particular to electrodynamics, and may not apply to gravitation. In other words, there may be values of Ampère's constant k other than $k = -1/2$.

---

[30] cf. R A R Tricker, *Early Electrodynamics: The First Law of Circulation* (Pergamon, Oxford, 1965), p 65.

Consider the case $k = -1$ with the same configuration as in Fig (4.12). The component of the force in the radial direction, $\hat{r}$, now is:

$$\hat{r} \cdot d^2 \vec{F}/dsds' = \frac{ii'}{r^2}\left(\frac{dz}{ds} - 2\frac{dr}{ds}\frac{z}{r}\right) = ii'\frac{d}{ds}\frac{z}{r^2}.$$

The projection of this force along the radial line connecting the two circuit elements vanishes when taken around a complete circuit. Now the invariant is simply $\cos\vartheta'/r$.

Not long after Ampère presented his final *memoire*, the German high school teacher, Grassmann proffered another force which gave the same total force as Ampère's. However, unlike Ampère's force which acts longitudinally along the line connecting the two circuit elements, Grassmann's force acts on only one of the current elements being proportional to the triple vector product,

$$id\vec{s} \wedge (i'ds' \wedge \hat{r}_{s'}),$$

or

$$i'ds' \wedge (id\vec{s} \wedge \hat{r}_s),$$

where $\hat{r}_s$ and $\hat{r}_{s'}$ are unit vectors in the $d\vec{s}$ and $d\vec{s'}$ directions, respectively.

Hence, whereas Ampère's law acts longitudinally, Grassmann's law acts transversely. We would expect that gravity, and any motion associated with it, act like Ampère's law, not Grassmann's. However, if accelerating masses emit gravitons, the putative analog of photons, then they should be emitted in the transverse directions to the motion. And just like the electric field which has velocity and acceleration components, so too, must gravity if it is to radiate as well as attract.

Whereas electrical fields can be shielded, gravitational fields know no barriers, according to Newton. However, according to Le Sage, the gravitational force is derived from a gravitational current, which if properly shielded, can result in gravitational attraction. This is known as 'shadow' theory. According to van Flandern, [31] this effect could be observed during eclipses, but the jury is still out for deliberation, and will probably be out indefinitely due to the opposition of mainstream cosmologists.

---

[31]T van Flandern, "Possible new properties of gravity," *Astrophysics & Space Sci* **244** (1996) 249-261.

The upshot of this all is that when dealing with gravitation, there is no *a priori* reason for keeping Ampère's choice of $k = -1/2$: what may be good for circuits may not be good for gravitating masses. In fact, the non-relativistic mechanical virial, (4.50), selects $k = 0$.

The eikonal equation,

$$n \frac{d}{ds'} \left(n \frac{dr}{ds} \right) = \tilde{F} \tag{4.52}$$

is a very beautiful equation since it extracts all the motional components of the force. For an index of refraction, $n = C/\sqrt{r}$, (4.52) gives Ampère's law, which describes light trajectories in a Coulomb field (cf. Eq (2.48)).

Rather, if we consider (4.52) to express a dynamical equilibrium, where the motional force is just balanced by the Coulomb field, then (4.52) is the condition for the vanishing of the Weber force. Whereas the eikonal equation (4.52) describes radial sectional curvature only [32], the Coulomb field has constant tangential curvature just like a sphere. The condition of dynamic equilibrium implies that the sectional curvatures are not independent, as we have seen on other occasions.

Ampère's law has an index of refraction $n = C/\sqrt{r}$. The eikonal (4.52) is:

$$\frac{1}{r} \frac{\partial^2 r}{\partial s \partial s'} - \frac{1}{2r^2} \frac{\partial r}{\partial s} \frac{\partial r}{\partial s'} = F_A. \tag{4.53}$$

Again, this shows that the eikonal is related to radial curvature. Tangential curvature must be added on, and when this is done, the result is Weber's force.

In general, we can write Ampère's law as:

$$d^2 F_A / ds ds' = \frac{ii'}{r^2} \left(r \frac{\partial^2 r}{\partial s \partial s'} + k \frac{\partial r}{\partial s} \frac{\partial r}{\partial s'} \right), \tag{4.54}$$

but, now, instead of current elements, $d\vec{s}$ and $d\vec{s}'$ we may think of them as two rays in media with different indices of refraction, where the radial vector, $\vec{r}$, is parallel to the normal between the two media.

---

[32] P Peterson, *Riemann Geometry*, 2nd ed (Springer, New York, 2006) §42.

Assuming spherical symmetry, the relative index of refraction,

$$ds'/ds = n(r),$$

will be a function only of r. Maxwell, [33] and his followers, made no qualms of distinguishing the two current elements when it came to replacing them by the time increment, dt, according to: [34]

$$e\dot{r} = c \cdot ids,$$

where c is "the number of units of statical electricity which are transmitted by the unit electric current in the unit of time." As such, c is a mere conversion factor that allows a static force to appear along side a motional one. But does the same conversion factor allow Coulomb's potential to be replaced by Newton's?

The cancellation of the different currents was based on Fechner's (incorrect) hypothesis that a single current consisted of plus and minus charges traveling in opposite directions with equal velocities. This is not the first time that a wrong assumption led to a correct result, but it was used as an arm to attack Weber's theory. The same can be said of Maxwell's mechanical concoctions that he used to derive his field equations. Miraculously, the edifice withstood the removal of the scaffolding.

Introducing the relative index of refraction into Ampère's law, (4.54) gives:

$$d^2 F_A / ds^2 = \frac{ii'}{r^2} \left\{ r \frac{d^2 r}{ds^2} + \left(k - \frac{r}{n} \frac{dn}{dr} \right) \left(\frac{dr}{ds} \right)^2 \right\}. \tag{4.55}$$

By requiring the longitudinal component of the force to vanish,

$$\frac{dn}{n} = k \frac{dr}{r}, \tag{4.56}$$

we get:

$$n = C r^k. \tag{4.57}$$

The force law, (4.55) thus reduces to the mechanical one:

$$\frac{d^2 r}{ds^2} = \frac{1}{ii'} \frac{d^2 F}{ds^2} r.$$

---

[33] *op. cit.* p 428 Eq (11).
[34] Maxwell, *op. cit.* §849 p 428.

The relative acceleration in the radial direction is:

$$\frac{d^2 r}{ds^2} - r\left(\frac{d\varphi}{ds}\right)^2 = -F_W \tag{4.58}$$

where F_W is the Weber force, the sum of the mechanical force, (4.51), and the Coulomb force, thanks to the condition on the index of refraction (4.56). For gravitation,

$$F_W = \gamma \frac{\mu}{c^2 r^2}\left(r\frac{d^2 r}{ds^2} + c^2\right),$$

we have replaced the currents ii' by μ/c^2. But not knowing the constant of proportionality we have designated it by γ, a coefficient to be determined by observation.

Transforming to $u = 1/r$, and changing the independent variable to φ, using $L = r^2 d\varphi/ds$ as the definition of the 'optical angular momentum', we get the equation of the orbit:

$$\left(1 + \gamma\frac{\mu}{c^2}u\right)\frac{d^2 u}{d\varphi^2} + u = \frac{\gamma\mu}{L^2},$$

where the only relativistic term is in the coefficient of the second derivative. The similarity with Gerber's equation for the advance of the perihelion of Mercury is unmistakable:

$$\left(1 + 6\frac{\mu}{c^2}u\right)\frac{d^2 u}{d\varphi^2} + u = \frac{\mu}{L^2} - 3\frac{\mu}{c^2}\left(\frac{du}{d\varphi}\right)^2. \tag{4.59}$$

Gerber, another German high school teacher like Grassmann before him, tried to fit (4.48) with a motional potential of the form:

$$V = -\frac{\mu}{r}\left(1 - \dot{r}^2/c^2\right)^{-2},$$

and came out with a value $\gamma = 3$, which gave the observed 43 arc-seconds per century for the advance of Mercury's perihelion. [35]

For time-retarded potentials the motional factor would appear to be first-order, so it behooved Gerber to find an explanation of why it should be squared. Roughly speaking, Gerber's potential requires a 'double-dose' of retardation.

---

[35] A K T Assis, in *Weber's Electrodynamics* (Kluwer, Dordrecht, 1994) p 204, says that the "discovery that a Weber's law applied to gravitation leads to a precession . . . has been rediscovered from time to time by many people." He then gives Gerber as an example. To the best of my knowledge Gerber knew the formula for the precession and needed to find a potential that would give him the right numerical factor. He was unaware that he derived Weber's force, and that it was the relative acceleration in that equation which gives the correct advance.

In 1917, the editor of the *Annalen der Physik* asked Gerber to prepare an expanded version of his 1902 paper so it would offer readers an alternative to Einstein's calculation using GR. As a backlash, it was deemed unsound and wrong through and through to match Gerber against Einstein. Yet, the same can be said of Einstein's modification of the orbital equation insofar as: [36]

> The relativity equation does not contain the slightest trace of 'warped space,' and the 'fourth dimension,' not the slightest trace of a new law of gravitation. It is an equation of Newtonian gravitation, purely and simply. . .

The equation GR comes up with is (2.78) to lowest order in α. There appears more similarity to Gerber's equation, (4.59), than to GR since the nonlinear term has been eliminated through the condition on the index of refraction, (4.56).

In fact, both equations effectively reduce to the harmonic oscillator form:

$$C\frac{d^2u}{d\varphi^2} + u = \frac{\gamma\mu}{L^2}.$$

The general solution to this equation is:

$$u = \frac{\gamma\mu}{L^2}\left(1 + \varepsilon\cos(\varphi/\sqrt{C})\right),$$

where the eccentricity, ε, appears as a constant of integration, and L^2/μ is the semi-latus rectum.

Moreover, if C were constant then $\gamma = 0$, and this would reduce to a Keplerian orbit. However, even for C constant but greater than 1, there will be a cumulative effect since φ must increase more than 2π in order to return to its initial position after a period of the motion.

Replacing r in $C = \sqrt{(1 + \gamma\mu/c^2r)}$ by its mean value via Kepler's third, $r = L^2/\mu$, realizing that φ has to increase by $2\pi\sqrt{C}$ in order to return to its initial position on the ellipse after a lapse of a period, we find the precession per period is $2\pi(C - 1)$,

---

[36]Poor, *op. cit.*

or (4.48) if we write in $\gamma = 3$. No classical theory can give the exact numerical factor, and who told Gerber that his potential is specific to the advance of the perihelion of Mercury? What about the perihelion shifts of the other planets which are well-known, and not all in the same direction, nor of the same magnitude.

It is only a coincidence that the linear term, $-6(\mu/c^2)d^2u/d\varphi^2$, in Gerber's equation, (2.78), and the nonlinear term, $3\alpha u^2$, in the GR equation, (2.78), both result in the same precessional term, $6\pi(\mu/cL)^2$. It then boils down to a matter of taste which one is considered more fundamental, seeing that neither can be applied to other perhelia.

Any nonlinear term in the equation of the orbit will lead to failure in closing the orbit. With a change in Ampère's constant, these nonlinear terms have been eliminated, which are related to the longitudinal component of the force.

However, it is wrong to conclude [37] that "Gerber did not in any way anticipate the two-body equation predicted by GR, let alone the field equations from which the relativistic equation of motion is derived." As we have repeatedly seen, GR is at most GO, and it cannot treat the two-body problems. Rather, it treats motion along geodesics whether they be heavenly bodies or rays of light.

Moreover, the field equations give no inkling into what causes the perihelion shifts, nor do they explain what causes the deflection of light by massive bodies. Neither does the vanishing of the Ricci tensor, since the Ricci tensor is an average of the sectional curvatures in a given direction. At best we are dealing with processes of 'gap-fitting" that will balance the books.

We now return to the derivation of the eikonal equation, (4.52). In doing so, we will learn a lot about the role of angular momentum in optics, and the different forms it takes in non-homogeneous systems.

We have seen that an index of refraction of the form $n = C/\sqrt{r}$ leads to Ampère's law in which the transverse speed, $\vec{\omega} \wedge \vec{r}$, is constant and implies induction , whereas an index of refraction of the form $n = C/r$ implies that the angular velocity, $\vec{\omega}$, or

---

[37] K Brown, *Reflections on Relativity* (lulu.com 2017)

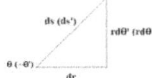

Figure 4.13: Right triangle illustrating the formula of Bouguer.

equivalently, the magnetic field, $\vec{B}$, is constant. This is the realm of magnetostatics, and, perhaps, gravitation.

Fermat's principle asserts that the true path will render the optical path,

$$I = \int n(r)ds = \int n(r)\sqrt{dr^1 + r^2 d\vartheta^2} = \int n(r)\sqrt{1 + r^2 \vartheta'^2}; dr,$$

an extremum, where the prime denotes differentiation with respect to r. Because φ is a cyclic coordinate we know that there is a first integral:

$$\frac{n(r)r^2\vartheta'}{\sqrt{1 + r^2 \vartheta'^2}} = L,$$

where L is the areal velocity which is conserved in a homogeneous medium of $n = 1$.

Now consider two phases, distinguished by a prime. Consider the right triangle in Fig (4.13).

Let ϑ be the angle between the tangent to the ray, ds, and the radius vector dr. The length of opposite side of the right-triangle will be $rd\vartheta'$, where ϑ' is the angle between the tangent to the ray, ds', and the radius vector, dr, in the second medium. The opposite side of the right-triangle will now be $rd\vartheta$.

Consequently, the areal velocities in the two media will be given by $L = \pm r \sin \vartheta'$, and $L' = \pm r \sin \vartheta$. The plus and minuses refer to the cases where ϑ' is either positive or negative. These are the well-known formulas of Bouguer.

We are now in a position where we can derive a relation between the gradient of the index of refraction and curvature, defined by:

$$K \equiv \frac{d}{ds}(\vartheta + \vartheta'). \tag{4.60}$$

Taking the logarithm of Bouguer's formula and then the derivative with respect to s result in:

$$\frac{1}{L'}\frac{dL'}{ds} = \left(\frac{\nabla n}{n} + \frac{1}{r}\right)\frac{dr}{ds} + \cot\vartheta'\frac{d\vartheta'}{ds} = 0.$$

Rearranging the formula shows how the gradient of the index of refraction is related to curvature (4.60):

$$\frac{\nabla n}{n}r\frac{d\vartheta'}{ds} = -K.$$

In other words, curvature is determined by the gradient of the index of refraction. For Ampère's law, or for any central potential varying as $1/r$, $\vartheta + 2\vartheta' = $ const. The important point to bear in mind is that L will be constant only in a homogeneous medium. When the media are no longer homogeneous, other conservation laws will apply.

If we consider:

$$\mathcal{T} = n(r)\frac{\partial r}{\partial s}\frac{\partial r}{\partial s'},$$

as a generalized kinetic energy, the Euler-Lagrange equation,

$$\frac{d}{ds'}\left(\frac{d\mathcal{T}}{\partial r/\partial s'}\right) - \frac{\partial \mathcal{L}}{\partial r} = n\frac{\partial^2 r}{\partial s\partial s'} = F/n, \tag{4.61}$$

will always coincide with the mechanical force, (4.51). In other words, the variational principle eliminates the longitudinal component of the force.

For $n = C/\sqrt{r}$, the force is Ampère's, written in the form:

$$F_A/C^2 = \frac{1}{r}\frac{\partial^2 r}{\partial s\partial s'} - \frac{1}{2r^2}\frac{\partial r}{\partial s}\frac{\partial r}{\partial s'}.$$

For this index of refraction, the angular velocity, $\vec{r} \wedge \vec{\omega}$, is constant. This confirms our earlier result. Taking the time derivative and identifying the angular velocity with the magnetic field, $\vec{B}$, the conservation condition reads:

$$\vec{v} \wedge \vec{B} + \vec{r} \wedge \frac{d\vec{B}}{dt} = 0. \tag{4.62}$$

The first term is the magnetic component in the Lorentz force, which, as we have observed, is actually the Grassmann force:

$$\vec{F}_G = \vec{v} \wedge \vec{B},$$

which pre-dates Lorentz by almost half a century.

Taking the vector product of $\vec{r}$ with Eq (4.62), and multiplying by $-\pi$, gives a torque:

$$\vec{\tau} = -\pi \left(\vec{r} \wedge \vec{F}_G \right) \equiv -\frac{d\vec{B}}{dt} \pi r^2.$$

The torque, τ, is the time-rate-of-change of the magnetic flux through an area πr^2, aka the electromotive force . Consequently, Ampère's law accounts for induction.

4.9 Converting elliptic into hyperbolic orbits

In this section we will derive the scattering angle for the deflection of light about a point target, which would correspond to the GR calculation if the impact parameter could be replaced by the radius of the massive body.

We have appreciated that the conformal transformation, $w = z^2$, converts Hooke's law for a harmonic oscillator into Newton's inverse square law. Here, we will show that it converts elliptic orbits into hyperbolic ones, and in doing so, derive the angle of scattering.

Consider a sphere of radius $r = \sqrt{2}$ which is filled with a material that gives a Hookean profile with index of refraction:

$$n^2 = 2 - r^2 \qquad r \leq \sqrt{2}. \tag{4.63}$$

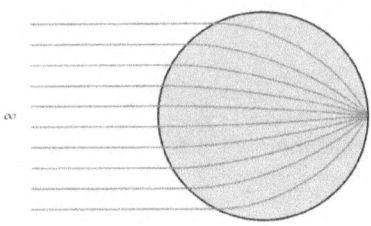

Figure 4.14: Luneburg lens where light rays come from ∞ enter the refractive index medium given by (4.63).

This is essentially a Luneburg lens where outside the lens, for $r > \sqrt{2}$, light propagates in straight lines with an index of refraction, $n = 1$. However, inside the lens, the rays no longer follow straight trajectories but, rather, bend into segments of an ellipse, as shown in Fig (4.14).

The eikonal equation is: [38]

$$\frac{d}{ds'}\left(n\frac{dr}{ds}\right) = c'^2 \nabla n. \tag{4.64}$$

This form of the eikonal is valid for vanishing force. Just consider Weber's law for which $n^2 = C/r$. The eikonal equation then takes the form:

$$\frac{d^2 r}{ds\,ds'} = \frac{1}{2}\left(\frac{c'^2 - dr/ds \cdot dr/ds'}{r}\right),$$

which is the condition for the vanishing of Weber's force, where c' is his constant.

We can consider the affine parameters to be equal and set $d\tau = ds/n$ so that (4.64) reduces to:

$$\frac{d^2 r}{d\tau^2} = \nabla n^2 = -r, \tag{4.65}$$

in natural units where $c' = \sqrt{2}$. This is the harmonic oscillator equation which we now wish to analyze.

---

[38] M Born & E Wolf, *Principles of Optics* (Macmillan, New York, 1959) p 121.

The solution to (4.65) can most easily be expressed in terms of complex variables:

$$z = \frac{a+b}{2}e^{i\tau} + \frac{a-b}{2}e^{-i\tau} = a\cos\tau + b\sin\tau.$$

The real amplitudes, a and b, have double meanings: a is both the length of the semi-major axis and the point of launch, and b is the length of the semi-minor axis, and the speed of launch.

The mapping, $z \mapsto z^2$ to the origin centered ellipse (i.e. a Hookean ellipse), shifts the origin to one of the foci:

$$z \mapsto z^2 = \frac{(a+b)^2}{4}e^{2i\tau} + \frac{(a-b)^2}{4}e^{-2i\tau} + \frac{a^2-b^2}{2}.$$

The first two terms correspond to the origin-centered ellipse with foci at $\pm(a^2 - b^2)/2$, while the last term translates the left-handed focus to the origin. But what about hyperbolic orbits?

Attraction and repulsion can be described in terms of the harmonic oscillator equations:

$$\frac{d^2z}{d\tau^2} + z = 0,$$

and

$$\frac{d^2z}{d\tau^2} - z = 0,$$

respectively. For the repulsive field, the solution to the latter equation is:

$$z = \frac{a+b}{2}e^{\tau} + \frac{a-b}{2}e^{-\tau} = a\cosh\tau + b\sinh\tau$$

where a is the point of launch and b is the speed of launch.

The mapping $z \mapsto z^2$ now transforms the pair of hyperbolas so they are origin-centered so that their foci are symmetrical with respect to the origin. But, the effect on its dual is more profound: it is still an inverse square law, but now it is repulsive .

This will have definite repercussions on the paths of the light rays. There are two parameters in the problem, Δ, the impact parameter, α, the Schwarzschild radius, which measures the magnitude of the gravitational force on light rays. They can be combined in the ratio: $\Xi = \Delta/\alpha$, which is a dimensionless constant.

The equation of the trajectory is (4.24), which in the present circumstance is:

$$\varphi = \int \frac{\Xi dr}{r\sqrt{r^2 n^2 - \Xi^2}},$$

where the index of refraction is given by the Luneburg lens, (4.63). Transforming to $x = r^2$, the integral becomes a standard one:

$$\int \frac{dx}{x\sqrt{ax^2 + bx + c}},$$

with a solution given by:

$$r^2 = \frac{\Xi^2}{1 - \varepsilon \sin 2\varphi}.$$

The eccentricity may be read off as:

$$\varepsilon = \sqrt{1 + \Xi^2} > 1,$$

which indicates a hyperbolic trajectory.

The angle of scattering, χ, is:

$$\cot \frac{\chi}{2} = \sqrt{\varepsilon^2 - 1} = \Xi.$$

for small values has $\cot(\chi/2) \approx 2/\chi$. The scattering angle is then given by:

$$\chi \approx \frac{2}{\Xi} = \frac{2\alpha}{\Delta},$$

which would be the GR result – if the impact parameter could be replaced by the radius of the deflecting mass. Negative gravity is associated with hyperbolic orbits.

5 The speed of gravity

5.1 Laplace on the speed of gravity

Again we turn to the *Principia*, this time the second edition published in 1713, and, in particular Prop 42 in Book 3. At the end of the discussion, Newton refers to Halley's observation that

> by combining the eclipse observations of the Babylonians with Albategeni's and today's, our Halley showed that the mean motion of the Moon when compared with the diurnal motion of the Earth, gradually accelerates. . .

The mention of the diurnal motion of the Earth left the door open to the possibility that Halley's observations could also be explained by a decrease in the Earth's rotation, as well as to an increase in the Moon's acceleration which would lead to a shorter lunar month than when the Babylonian made their observation thousands of years ago.

Halley's explanation that the presence of a luminiferous ether was the cause of the apparent secular acceleration of the Moon's orbit to become increasingly smaller. Laplace in Vol IV, Book X, Ch VII of *Mecanique Celeste* quantified this as considering gravity as a type of fluid moving away from its source that would propagate at a velocity v_G. The Moon being bombarded by such particles would certainly slow down, and gravitational attraction could be explained on the basis that whenever, or wherever, there is a shielding of such particles by other masses, there is a diminution in their number, and, consequently, an attraction that depends on the mass of the shielding body.

The tiny flux of particles would act in a similar way that a radiation pressure acts in electromagnetism. Le Sage, an eighteenth century physicist, is usually credited with this 'shadow' theory of gravity in which the presence of a neighboring mass shields the mass from the influx of particles thereby creating an attractive force. Moreover, as in the sticky bead argument of gravitational waves to be discussed in the next chapter, the absorption of these particles would cause bodies to heat up and re-radiate the thermal energy so as to maintain a state of dynamic thermal equilibrium.

Van Flandern [1] has claimed that planets do radiate more heat back into space than they absorb from the Sun, so that this could be partially due to the absorption of these tiny particles. If gravity traveled at the speed of light, or any other finite speed, there would be a time lapse between the emission of gravity from the Sun until its arrival on Earth. If it does travel at the speed of light there would be a 500 seconds delay so that when it arrives on Earth the position of the Sun would have moved. This difference amounts to around 20 seconds of an arc.

That is, from the perspective of the Sun, the Earth is moving at a speed of 10^{-4} c, so that light from the Sun strikes the Earth at a slightly forward angle of 10^{-4} radians, the ratio of the speed of the Earth to that of light. This is the 20 arc seconds, and the angular displacement is called *aberration*. One of the most convincing arguments that Van Flandern has used against GWs propagating at the speed of light is that this gravitational aberration has never been observed.

With 20/20 hindsight, Laplace's expression for the rate of change of the period of the Moon can be derived in a very simple manner. From Kepler's III, we have:

$$-\frac{\dot{\omega}}{\omega} = \frac{\dot{P}}{P} = \frac{3}{2}\frac{\dot{a}}{a}, \tag{5.1}$$

where P is the period.

Suppose that v_G is the speed of propagation of the gravitational force. If a_0 is the length of the semi-major axis at time t_0, then ordinary perturbation theory of celestial mechanics [2] gives the secular variation of the semi-major axis at any other time t as:

$$a^2(t) = a_0^2 + 4\frac{\mu}{v_G}(t - t_0), \tag{5.2}$$

where μ is the gravitational parameter, the product of the gravitational constant and the total mass of the system.

---

[1] T Van Flandern, "Possible new properties of gravity," *Astrophysics and Space Science* **244**(1996) 249-261.
[2] J M A Danby, *Fundamentals of Celestial Mechanics* (Macmillan, New York, 1962) ch 11.

Taking the time derivative of (5.2) and introducing it into (5.1) give:

$$\dot{P} = 6\pi \frac{a\omega}{v_G},\tag{5.3}$$

where $\omega = 2\pi/P$, and $\mu/a^2 = a\omega^2$ have been used. With $\dot{\omega} = 20$ arc seconds, the speed of gravity, from Laplace's equation (5.3) is found to be:

$$v_G \approx 7.5 \times 10^6 c.$$

In his treatise, Laplace concluded "we must suppose that the gravitating fluid has a velocity which is at least a hundred millions of times greater than that of light." We should modify that to 7 million, but what does that matter?

Using modern astronomical observations, Van Flandern upped the ante to $2 \times 10^{10}c$. This is infinite for all intent and purposes.

Yet, if we look to the advance of the perihelion of Mercury, and if it is a gravitational wave, then it travels at a speed slower than that of light by a factor of $\sqrt{3}$. This is better understood as light passing through a denser medium due to a varying gravitational potential.

5.2 Gravitational aberration

Conventional wisdom claims that accelerating masses emit gravitational radiation in just the same way that accelerating charges emit electromagnetic radiation. And if gravitational radiation propagates at the speed of light it will be aberrated just like electromagnetic radiation.

Consider a star A orbiting about a fixed star B, as shown in Fig (5.1). The true position is A_t, while its linearly extrapolated position from A_r is A_c. With a as the radius of the orbit, the angle at B which A makes in time a/c is θ. The angle ψ is the angle at B formed from the linearly extrapolated position A_c and its true position A_t. By

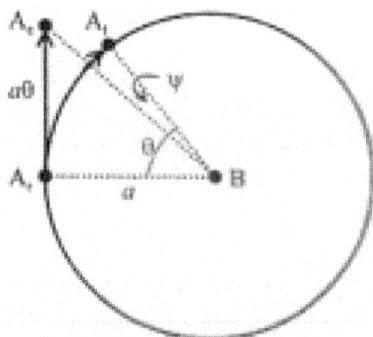

Figure 5.1: A comparison of the star's true position A_I with its linearly extrapolated position A_c. Taken from Van Flandern, "The speed of gravity: What the experiments say." *op. cit.*

construction, the distances from the past position, A_r, to the linearly extrapolated position, A_c, and the true position, A_t, are both equal to $a\theta$.

Although the distance between A_c and A_t is small, ψ will cause the extrapolated position to feel a tangential force that will continue to increase as A makes its way around the orbit. By simple geometry:

$$\tan(\theta - \psi) = \theta.$$

Inverting this expression and considering θ to be a small angle, $\tan^{-1}\theta$ can be developed in a series in θ beginning with:

$$\psi = \frac{1}{3}\theta^3 - \frac{1}{5}\theta^5 + \cdots \qquad (5.4)$$

At this point, it is worth emphasizing that we are not allowed to select any order in aberration we please in (5.4) without showing that all lower terms are zero. This is relevant to the Peters-Mathew expression for the luminosity which we will discuss in the next chapter. Whereas black body radiation is third-order, their expression is seventh-order, and it doesn't explain why or how all lower terms do not contribute to the luminosity.

Now θ is the product of the angular speed, $\omega = 2\pi/P$ and the light time a/c, or $a\omega/c$. The radial acceleration is $a_r = \mu/a^2$, so, to lowest order, the tangential acceleration experienced by A is:

$$a_t = \frac{\mu}{a^2}\psi = \frac{1}{3}\frac{\mu}{a^2}\left(\frac{a\omega}{c}\right)^3 - \frac{1}{5}\left(\frac{a\omega}{c}\right)^5 + \cdots \tag{5.5}$$

And instead of the linear relation, (5.1), for the rate of increase in the period, we now have the cubic relation:

$$\dot{P} = 2\pi\left(\frac{a\omega}{c}\right)^3 + \cdots . \tag{5.6}$$

Brown, in his book *Reflections on Relativity*, launched a scathing attack on Van Flandern [3] saying that Laplace's result, (5.3), is not (5.6), and, therefore it was "utterly absurd," and complete "gibberish." Since it will be important for our discussion of the luminosity expression that was used by Taylor & Hulse, in the analysis of gravitational radiation from a dissipative binary, and LIGO's analysis of GWs, both of which we shall discuss in the next chapter, it is more than advantageous to dwell on the distinction between orders of aberration.

In order to do so, we return to our discussion of the radius of curvature in Newton's proof of his inverse square law that we discussed in §1.1. Kepler's III: $a^3\omega^2 = $ const. can be written as $av^2 = $ const., by introducing the angular velocity, $v = a\omega$. Now, this means that the sine of aberration angle, α, in the expression of the angular momentum, (2.2), $L = av\sin\alpha$, must be proportional to the velocity, and we wrote it as $\sin\alpha = \Delta/nr$, where Δ was the impact parameter, and $n = c/v$, the index of refraction. From $av^2 = $ const. we deduce that:

$$\frac{\dot{a}}{a} = -2\frac{\dot{v}}{v}, \tag{5.7}$$

and for $\dot{a}$ we use (5.2) to obtain $\dot{a} = 2\mu/v_G a$. Then from (5.7) with $\dot{v} = a_t$, the tangential acceleration, we find:

$$a_t = -\frac{\mu}{a^2}\frac{v}{v_G}, \tag{5.8}$$

which is Laplace's starting point, without having to make any assumption that it is a first-order aberration term. [4]

[3]T Van Flandern, "The speed of gravity—What the experiments say," *Phys Lett A* **250** (1998) 1-11.
[4]K T McDonald, "Laplace and the speed of gravity".

Eliminating $\dot{a}$ in favor of $\dot{P}$, and using Newton's condition of uniform acceleration, $v = 2\pi a/P$, we easily arrive at Laplace's equation — but this time with a minus sign. The negative sign means that the period is getting shorter, as Halley believed. In terms of Le Sage's 'shadow' theory the sign depends on whether the gravitational particles are emanating from the source, as in 'shadow' theory, or directed toward the source, which would be its sink. In the former case, the period would be increasing.

Later Darwin predicted the opposite to what Halley had supposed; that is, tidal forces would disrupt and lead to an increase of the Moon's period, thereby lengthening the Earth's day that would be caused by the increasing distance between the Earth and the Moon.

So Laplace's result is first-order aberration, based upon secular perturbation, (5.2), and Kepler's laws. What about the third-order aberrational effect, (5.6)?

Recall that Newton wrote his central force as, $F_c \sin \alpha = v^2/\varrho$. With the expressions for curvature (2.112) and slope, it is easy to see that the curvature is related to the semi-major axis, a, by:

$$\varrho = a/\sin\alpha. \tag{5.9}$$

If a were constant, (5.9) would tell us we have a first-order aberrational effect. But, Newton's equation:

$$\varrho \sin^3 \alpha = L^2/F_c a^2 \tag{5.10}$$

tells a different story. If Newton's inverse-square law holds, then we have a third-order aberrational effect:

$$\varrho \sin^3 \alpha = \text{const.}$$

Aberration will thus cause the curvature, $1/\varrho^2$ to increase.

In other words, if a were constant, the radius of curvature would undergo a first-order aberration effect, (5.9). But, due to the variation of a in time, (5.2), it will become a third-order effect, with the consequence of ensuring Newton's inverse square law holds.

5.3 Geodesics with definite metrics

Consider a curve in parametric form, $\alpha = h(x(\tau))$, where τ is 'time.' The velocity of the curve is:

$$\dot{\alpha}_k = h_{k,j}\dot{x}_j, \tag{5.11}$$

where the comma stands for differentiation, and there is a sum on repeated indices.

For geodesic motion (5.11) must maintain its constant magnitude, but it is allowed to change direction. Multiplying (5.11) by $\dot{\alpha}_k$, gives:

$$\dot{\alpha}_k\dot{\alpha}_k = h_{k,i}\dot{x}^i h_{k,j}\dot{x}^j = g_{ij}\dot{x}^i\dot{x}^j,$$

where

$$g_{ij} = h_{k,i}h_{k,j} \tag{5.12}$$

defines the metric. The condition that the velocity,

$$V = \sqrt{\left(g^{ij}\dot{x}_i\dot{x}_j\right)}, \tag{5.13}$$

be constant will yield the geodesic equations, and characterize them as being the shortest path through given endpoints at times τ_1 and τ_2,

$$\int_{\tau_1}^{\tau_2} V(x,\dot{x})d\tau = \text{extremum}. \tag{5.14}$$

The condition is:

$$V_{x_i} - (V_{\dot{x}_i x_k}\dot{x}_k + V_{\dot{x}_i \dot{x}_k}\ddot{x}_k = 0).$$

In order to get a unique solution, the assumption that (5.13) remain constant is added: [5]

$$\dot{V} = 0, \tag{5.15}$$

which says that

$$V_{x_k}\dot{x}_k + V_{\dot{x}_k}\ddot{x}_k = 0.$$

---

[5] P Pokorny, "Geodesics Revisited," *CMSIM* (2012) 281-298.

The resulting equations can either be written as:

$$2\frac{d}{d\tau}\left(g^{ij}x_j\right) - \frac{\partial g^{ij}}{\partial x_j}\frac{dx_i}{d\tau}\frac{dx_j}{d\tau} = 0, \tag{5.16}$$

or

$$\ddot{x}^k + \Gamma^k_{ij}\dot{x}_i\dot{x}_j = 0, \tag{5.17}$$

if we introduce the Christoffel symbol:

$$\Gamma^k_{ij} = \frac{1}{2}g^{kl}\left(g_{il,j} + g_{lj,i} - g_{ij,n}\right) \tag{5.18}$$

where g^{ij} is an element of the inverse matrix to g_{ij}.

It is crucial to bear in mind that the Christoffel symbols do not account for curvature, for that, you need second derivatives of the metric, or, equivalently, the derivative of the Christoffel symbol. And it is for this reason that the gravitational tensor of Einstein is only a 'pseudo-'tensor. By a change of coordinates, the connection coefficients Γ^k_{ij} can be made to vanish, and with them the gravitational, pseudotensor. .

The geodesic equations, (5.17) can also be arrived at by considering the second derivative of the path, α. It will be shown that the requirement is for the acceleration vanish in the tangent plane to the surface where our geodesic lies. Only the perpendicular component to the surface can subsist, as we now show. The acceleration is:

$$\ddot{\alpha}_i = h_{i,jk}\dot{x}_j\dot{x}_k + h_{i,j}\ddot{x}_j. \tag{5.19}$$

The last term in (5.19) is not the particle's acceleration. Rather, the left-hand side is, and if we impose the condition that the tangential component of the acceleration be zero,

$$\ddot{\alpha}_k h_{k,j} = 0, \qquad \forall j \tag{5.20}$$

we come out precisely with the geodesic equation. Thus, the geodesic equation can rightly be interpreted as guaranteeing that the only component of acceleration which can subsist is the normal component, and this leaves the velocity constant along the path; hence, the name 'geodesic.'

5.4 Fictitious accelerations and geodesics

The distinction between a^i and $\ddot{x}^i$ depends on whether we are in a non-inertial system or not. Fictitious, or pseudo-forces, arise in non-inertial systems such as rotating frames, the prototype of which is the centrifugal force, $a = v^2/r$, acting to propel a body away from its center of rotation. In terms of the curvature of the coordinate system which the rotating frame introduces, the total acceleration is comprised of a series of terms given in terms of the Christoffel symbols,

$$a^i = \ddot{x}^i + \Gamma^I_{jk}\dot{x}^j x_k + 2\Gamma^i_{j0}\dot{x}^j + \Gamma^i_{00},$$

where, on account of $4D$ space-time, gravity has been included as a fictitious acceleration in the last term.

Just consider the winds about a rotating body, like the earth. The centrifugal acceleration, $\Gamma^i_{jk}\dot{x}_j\dot{x}_k$, affects the curvature of the winds, while the Coriolis acceleration , $2\Gamma^i_{j0}\dot{X}_j$, modifies the direction of the winds. Moreover, if the rotation happens to be non-uniform, an additional fictitious acceleration will sneak in, i.e. the so-called Euler acceleration, which is the moment of the time rate of change of the angular velocity vector, ω^i. Since this is a completely tangential acceleration, it should have no effect upon a radial directed gravitational acceleration.

Gravitational acceleration is considered as a fictitious acceleration in general relativity because of Einstein's equivalence principle. In a frame accelerating with the same magnitude but in the opposite direction, gravitational acceleration can be made to disappear. This is in contrast to the transition from a non-inertial frame to an inertial where the fictitious accelerations vanish. However, the former is an accelerating frame while the latter is a constant velocity one.

Consider the Poincarè disc model whose metric is

$$\dot{\tau}^2 = \frac{\dot{r}^2 + (r\omega)^2}{(1 - (r\omega)^2)^2} = 2T.$$

If gravity is present, we can either consider it as an actual force, or a fictitious force. indexgravity, real and fictitious One-half of this expression is the Lagrangian, and if the gravitational force is "real", the Lagrangian would be modified to read $\mathcal{L} = T - \phi$, where $\phi = \mu/r$ is the gravitational energy. In contrast, if gravity were a fictitious force, kinetic energy would be given by

$$T = \frac{1}{2}\frac{\dot{r}^2 + (r\omega)^2}{(1 - (r\omega)^2 - 2\mu/r)^2},$$

As such, this would be conformally equivalent to the exterior Schwarzschild solution. [6]

The Euler-Lagrange equations give the geodesic equation,

$$\ddot{r} - r\omega^2 - \frac{r\omega^2 - 2\mu/r^2}{(1 - (r\omega)^2 - 2\mu/r)^2} \approx \ddot{r} - r\omega^2 + \Gamma^r_{rr}\dot{r}^2 = 0.$$

where in the second expression we have left out terms of ω^2, and higher, therefore being able to cast this term as a fictitious force in terms of the diagonal Christoffel symbol. As such it gives an additional contribution to the centrifugal acceleration, beyond that given by $-r\omega^2$. However, the curvature is no longer constant.

The additional contribution is definitely non-Euclidean; it modifies the boundary of the Poincarè disc. Rather, had we considered it an actual force, it would act on the particle in the disc as an impressed force, and not a mere modification of the disc's rim. Likewise, the Coriolis acceleration enters when we transform the rotational velocity to a moving frame, $v \to \dot{r} - \omega \wedge r$. Again, this fictitious force shows up as $\Gamma^r_{r0}\dot{r}$, and not as an impressed force in the geodesic equation. For if it weren't so, it would nullify the validity of there being a geodesic equation to begin with.

If we were to bury gravity as an impressed force, there would be no such thing as a Poisson equation, derivable from a Lagrangian

$$\mathcal{L} = -\rho\phi - \frac{1}{8\pi G}(\nabla\phi)^2,$$

---

[6]C. Moller, *The Theory of Relativity*, p. 325, Eq (79).

where ρ is the mass density. Variation with respect to ϕ gives the Poisson equation

$$\nabla\left(\frac{\partial\mathcal{L}}{\partial\nabla\phi}\right) = \frac{\partial\mathcal{L}}{\partial\phi}$$
$$\frac{1}{4\pi G}\nabla^2\phi = \rho.$$

Since Einstein's equations can be considered generalization of the Poisson equation, Einstein, himself, was considering ϕ as a potential, and its derivative as a fictitious force.

Nevertheless, fictitious forces have played a central role in general relativity which assumes the existence of a "gravito-magnetic" force—the gravitational analog of the magnetic field. For this is the only way to get traverse waves from two orthogonal oriented fields. However, the gravitational analog of Faraday's assumption that a moving conductor in a stationary magnetic field is the same as a stationary current in a varying magnetic field is non-existent!

Ampère's and Faraday's laws are related to fictitious accelerations: The former is a projection of the sum of a radial contraction in the direction of motion and the Coriolis acceleration, while the latter is an example of Euler's acceleration for a non-uniform magnetic field. In the absence of these fictitious acceleration there can be no propagation of EM waves.

It has been assumed that the Lorentz acceleration is really a Coriolis acceleration. However, whereas the former subsists in an inertial frame, the latter does not. Lorentz's acceleration uses the velocity of the moving charge $\dot{r}$, whereas the Coriolis acceleration employs the relative velocity, v. The two are related by $v \cdot \hat{r} = \dot{r}$, where $\hat{r}$ is a unit vector pointing in the radial direction.

Now, if the total electric field consists of a contraction of the field in the radial direction and the Coriolis acceleration,

$$\tilde{E} = E\left(1 - \frac{\dot{r}^2}{c^2} - 2v \wedge \omega\right),$$

where the angular speed is

$$\omega = \frac{v \wedge \hat{r}}{r^2 c^2},$$

then the electric field is

$$\tilde{E} = E\left(1 - \frac{\dot{r}^2 - 2(v \wedge v \wedge \hat{r})}{c^2}\right).$$

Using the vector identity, $v \wedge v \wedge \hat{r} = (v \cdot \hat{r})v - v^2 \hat{r}$, and taking the projection of the field in the radial direction give

$$\hat{r} \cdot \tilde{E} = |E|\left(1 + \frac{2v^2 - 3\dot{r}^2}{c^2}\right)$$

which is precisely Coulomb's and Ampère's laws. Thus, Ampère used a relativistic correction in the direction of the motion and a Coriolis accelerated both projected in the radial direction which is deduced, unknowingly, from the study of charges in motion!

What Ampère actually considered were two conducting elements, s and s', separated by a distance r. The angles that each of these elements make with respect to the line $\cos\theta = dr/ds$, and $\cos\theta' = dr/ds'$. In the limit, both of these will merge into the particle velocity, $\dot{r}$. Alternatively, the angle between the two current elements, ε, without regard to the distance between them, is v, where $ee'v^2 = jj'\cos\phi$, where $j = ids$ is the current in element ds. Thus, Ampère's law will take the more familiar form

$$\hat{r} \cdot \tilde{E} = \frac{j'j}{r^2}\left(2\cos\varepsilon - 3\cos\theta\cos\theta'\right).$$

5.4.1 Unmasking the Lense-Thirring effect and the de Sitter precession

General relativity, through its "gravito-magnetism", has claimed two phenomena of its own: the Lense-Thirring (LT) effect, and the de Sitter (dS) precession. According to GR, the LT phenomenon is "frame-dragging", and dS precession is the result of the non-commutivity of sequential, non-aligned Lorentz transformations, and the gravitational acceleration that causes space-time to curve. In order to do these phenomena full justices they must be analyzed within the framework of the Kerr metric.

LT precession arises from a gyroscope orbiting a rotating central body, and it is even present when the gyroscope is at rest. The precession at the poles is parallel to the rotation of the central body whereas it is anti-parallel at the equator. This has given rise to the colorful, if misleading expression, that the rotation of the central body "drags" the metric along with it. At the poles, the metric tends to rotate in the same direction as the central body whereas the "dragging" of the metric falls off with increasing distance at the equator. If the gyroscope is oriented perpendicularly with respect to the equator, the side way motion from the central body will be dragged less than it would be if were closer to the body. Thus, the spin is predicted to precess in the opposite direction to the direction of rotation of the central body.

Additionally, the dS precession is the gravitational analog of the Biot-Savart law, where the creation of the magnetic field due to the flow of charge is replaced by the flow of matter which creates a gravitational field. However, the latter can exist even without the "flow" of matter.

The vector potential in the LT precession is analogous to that of a vortex,

$$2\pi A = \frac{GI}{c^2}\frac{\omega}{r},$$

where I is the moment of inertia of the rotating body and ω is its angular speed. In analogy with vortex motion the vector potential would be replaced by the vortex strength, $\omega/r \cdot S$, where S is the volume, the area times the height, $h = GM/c^2$, the mass coming from the moment of inertia, $I = Mr^2$, neglecting numerical coefficients that depend on the shape of the body.

The velocity of vortex flow is the curl of the vector potential,

$$v = \nabla \wedge A = \frac{GI}{c^2}\frac{\omega \wedge r}{r^3}.$$

The gravito-magnetic field appears as the curl of this velocity,

$$H = \nabla \wedge v = \frac{GI}{c^2 r^3}\left(3(\omega \cdot \hat{r})\hat{r} - \omega\right).$$

It is clear from the demonstration that H is is $3D$, so it is a little perplexing to see it used to describe a black hole, which from all conventional accounts is $2D$. [7] Thus,

---

[7] see, K Thorne, *The Membrane Paradigm*

the he LT and dS precessions are nothing more than fictitious forces parading as new phenomena discovered by GR.

5.4.2 Derivation of Gerber's equation

Newton's equation for an ellipse, as well as Cotes' spiral, can be derived by differentiating the path (4.24) in time, and using the conservation of angular momentum, written in the form of a geodesic equation:

$$r\ddot{\varphi} + 2\dot{r}\dot{\phi} = 0. \tag{5.21}$$

We then obtain:

$$\ddot{r} - r\dot{\varphi}^2 = \nabla n^2. \tag{5.22}$$

For a spatially isotropic, but inhomogeneous, medium, the index of refraction can depend only on r, and it should be a decreasing function of it. For instance, for the index of refraction given by an Eaton lens, (2.48), Eq (5.22) gives ellipses. For $n^2 \propto 1/r^3$ there results Cotes's spiral, and so on.

However, the index of refraction cannot be a function of the speed of the wave, if the motion is to remain geodesic. A case in point is Gerber's equation, where the index of refraction is given by:

$$n^2 = \frac{c^2 - \dot{r}^2}{r}. \tag{5.23}$$

For introducing (5.23) into the right-hand side of (5.22) there results:

$$\ddot{r} - r\dot{\varphi}^2 = -\left(\frac{c^2}{r^2} + \frac{\dot{r}^2}{r^2} - 2\frac{\ddot{r}}{r}\right), \tag{5.24}$$

which is Gerber's equation to within an undetermined multiplicative constant. The right-hand side also happens to be the Weber force! So there is more to the Gerber potential than meets the eye. It was brought in here, though, as an example of a non-geodesic equation of motion. In contrast, Einstein's modification of the path trajectory, (2.78), does not contain accelerations, and so remains a geodesic equation.

Assis, [8] sets the right-hand side of (5.24) equal to the Weber force, but does not give any reason for doing so. Roseveare, [9] gives a lengthy criticism of Gerber's work, but

---

[8] A K T Assis, *Weber's Electrodynamics*(Kluwer, Dordrecht, 1994) p 204 Eq (7.40).
[9] N T Roseveare, *Mercury's Perihelion from Le Verrier to Einstein* (Oxford U P, Oxford, 1982).

does not realize that the right-hand side of (5.24) is the Weber force if the speed of propagation is $c/\sqrt{3}$. He pinpoints the cause of the advance of the perihelion to the acceleration term on the right-hand side, but the numerical coefficient in the perihelion advance could have been fixed *a priori* by demanding that the right-hand side be exactly given by the Weber force with a speed $c/\sqrt{3}$.

5.4.3 Snell's law & geodesics in non-euclidean geometries

Examples of geodesics abound from geometrical optics and non-Euclidean geometries of constant curvature. The analogy with geodesics is what probably motivated Einstein to introduce geodesics in the first place.

Light rays also follow geodesics, and Snell's law is its manifestation. Consider a light beam in a plane with the y-axis normal to the surface of a medium consisting of two media with differing index of refraction. The relative index of refraction, $n = n(y)$, will only be a function of the distance from the surface.

The metric:

$$c^2 dt^2 = n^2(y)\left(dx^2 + dy^2\right) \tag{5.25}$$

is compatible with the geodesic equation, (5.17). For $x_1 = x$ and $x_2 = y$, the geodesic equations read:

$$\ddot{x} + 2\nabla \ln n \; \dot{x}\dot{y} = 0,$$

and

$$\ddot{y} + \nabla \ln n \left(\dot{y}\dot{y} - \dot{x}\dot{x}\right) = 0.$$

Multiplying the first of these equations by $\dot{y}$, and the second by $\dot{x}$, and subtracting one from the other give:

$$\ddot{y}\dot{y} - \ddot{x}\dot{x} = \nabla \ln n \; \dot{x}\left(\dot{x}^2 + \dot{y}^2\right).$$

By multiplying through by $2\dot{y}n^2/\dot{x}^3$ [10] leads to a total differential in time:

$$\frac{d}{dt}\left(\frac{1 + (dy/dx)^2}{n^2}\right) = 0,$$

---

[10]Pokorny, *op. cit.*

implying that there exists a first integral of the motion:

$$\frac{1 + (dy/dx)^2}{n^2} = \text{const.}$$

If ϑ is the angle formed from the light beam and the normal to the surface, $dx/dy = \tan\vartheta$, the integral of the motion can be written as:

$$n \sin \vartheta = \text{const.,} \tag{5.26}$$

which is none other than Snell's law.

The metric (5.25) indexconformal metric is conformal insofar as lines are distorted but angles remain the same. One way of getting a new geometrical structure is to distort an old one.

A case in point is a system with an index of refraction, $n = 1/y$, for then the metric (5.25) becomes:

$$ds^2 = \frac{dx^2 + dy^2}{y^2},$$

which is the metric for the Poincarè half-plane model of a hyperbolic geometry with constant curvature. It satisfies a geodesic equation with a Christoffel symbol, $\Gamma^x_{xy} = 2\nabla \ln n$. Snell's law, (5.26) now reads

$$\sin \vartheta / y = C,$$

which is the equation of a circle of radius $R = 1/C$. Instead of being straight lines, as they are in Euclidean space, the geodesics are now semi-circles that cut the x-axis orthogonally, as shown in Fig (5.2).

In the limit of infinite radius, the geodesics becomes straight lines, $x = \text{const.}$

Test particles moving on semi-circles are not in 'free-fall', and those moving on a straight line are not in a '*local* geodesic coordinate frame.' To read otherwise is a complete distortion of what geodesics represent; they certainly do not represent accelerations. We will return to these points shortly.

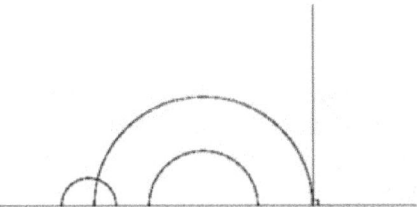

Figure 5.2: Geodesics of the Poincarè half-plane model are semi-circles or straight lines parallel to the y-axis.

5.4.4 Derivation of Ampère's law

The derivation of Ampère's law as a geodesic equation offers a new twist with constant angular velocity replacing the conservation of angular momentum.

We begin, again, with the conformal metric, (5.25), except now we write it in polar coordinates:

$$V^2 = n^2(r) \left(\dot{r}^2 + r^2 \dot{\varphi}^2 \right). \tag{5.27}$$

The radial geodesic equation is:

$$\ddot{r} - r\dot{\varphi}^2 + \nabla \ln n \left(\dot{r}^2 - r^2 \dot{\varphi}^2 \right) = 0. \tag{5.28}$$

The angular geodesic equation leads to the conservation law:

$$n^2 r^2 \dot{\varphi} = \text{const}. \tag{5.29}$$

For a transparent medium (5.29) gives the conservation of angular momentum. Rather, for an index of refraction like the Eaton lens, (2.48) but without the constant term, (5.29) gives the constraint of constant angular velocity. Calling this V, the radial geodesic, (5.28) becomes:

$$\ddot{r} = \frac{\dot{r}^2 + V^2}{2r} > 0, \tag{5.30}$$

showing that it is repulsive. Multiplying through by ee'/r, and rearranging give:

$$\frac{ee'}{r^2} \left\{ r\ddot{r} - \frac{1}{2}\dot{r}^2 \right\} = \frac{ee'}{r^2} V^2, \tag{5.31}$$

which is Ampère's law for the force that arises from the motion of two charges e and e' which repel each other.

It seems strange that this is geodesic motion at constant angular velocity, but we can show that it corresponds to a conic section when transformed to φ as the independent variable under the condition $V = $ const. A further transform to $u = 1/r$ gives the equation of the path trajectory as:

$$2\frac{d^2u}{d\varphi^2} + u = \frac{1}{u}\left(\frac{du}{d\varphi}\right)^2,$$

whose solution is a parabola:

$$u = A(1 + \cos(\varphi - \varphi_0)),$$

where A and φ_0 are two constants of integration.

5.5 Geodesics with indefinite metrics

Einstein implanted geodesic motion into GR saying that

> The field equations of gravitation... which assert that a gravitating particle moves in a geodesic line. This constitutes a hypothetical translation of Galileo's law of inertia to the case of the existence of a 'genuine' gravitational field.

With the introduction of an indefinite metric, it is advantageous to introduce a velocity 4-vector v^μ, where if the affine parameter is the proper time τ then

$$g_{\mu\nu}v^\mu v^\nu = g_{\mu\nu}\frac{dx^\mu}{d\tau}\frac{dx^\nu}{d\tau} = c^2. \tag{5.32}$$

Constant velocity implies:

$$g_{\mu\nu}v^\nu\left(\frac{dv^\mu}{d\tau} + \Gamma^\mu_{\alpha\beta}v^\alpha v^\beta\right) = g_{\mu\nu}v^\nu a^\mu = 0,$$

where a^μ is referred to as *proper* acceleration.

One reason for introducing the velocity 4-vector is that even if all the spatial components of this vector are zero, it does not mean that the components of the 4-acceleration are zero. GR expresses this as even if the body is at rest in such a reference frame, the body is still experiencing acceleration, and something must be causing that acceleration.

This is how gravity is perceived, but it is a 'force' that must be causing that acceleration. In this way, gravity has been relegated to the time component of the metric. In other words, *it is no longer a force, but it is still a force because it causes acceleration.*

Even in Euclidean geometry, what is constant speed in Cartesian coordinates, $\ddot{x} = 0$ does not carry over to polar coordinates. There, as we have seen, the condition of constant speed is given by the geodesic equation (5.17) where $\Gamma^r_{\varphi\varphi} = -r$, and $\Gamma^\varphi_{r\varphi} = \Gamma^\varphi_{\varphi r} = -1/r$ in the plane $\vartheta = \pi/2$.

The geodesic equations appear much less useful in GR than the first integrals themselves. Recalling Danby's remark, that a first-order differential equation is to be preferred to a second-order one when the field is conservative. The only advantage of the geodesic equations is when the force can be expressed as a power series in the inverse radial coordinate. It will therefore be advantageous to work with (5.32) directly, when proper time is replaced by an affine parameter. We thus return to the discussion of gravitational repulsion, and the derivation of the orbit from the condition of constant velocity, which has one less constant of integration to deal with.

5.5.1 The Case Against Gravitational Repulsion

We have discussed already how a distant observer, using a gravimeter, can claim that negative gravitational repulsion can occur anywhere in the Schwarzschild field. We will now show that the error in the deduction of a repulsive gravitational field is due to the lack of application of the geodesic constraint, (5.11).

Recall that the equation of motion used by McGruder, [11] who references Weinberg [12] is:

$$\ddot{r} + \frac{A'}{2A}\dot{r}^2 - \frac{r}{A}\dot{\vartheta}^2 - \frac{r^2 \sin\vartheta}{A}\dot{\varphi}^2 + \frac{B'}{2A} = 0, \tag{5.33}$$

where the dot now means differentiation with respect to an affine parameter, and $AB = 1$, with $B = (1 - \alpha/r)$.

To (5.33), we now apply the condition that the square of the velocity be constant:

$$V^2 = B\dot{t}^2 - A\dot{r}^2 - r^2\dot{\vartheta}^2 - r^2 \sin^2\vartheta\,\dot{\varphi}^2, \tag{5.34}$$

be constant. Without loss of generality, we restrict the motion to lie in the equatorial plane, $\vartheta = \pi/2$. By eliminating $\dot{t}^2$ between (5.33) and (5.34) we get the equation of motion in polar form as:

$$\ddot{r} - r\dot{\vartheta}^2 \left[\frac{1}{A} - \frac{1}{2}B'r\right] = -\frac{1}{2}B'V^2 - \frac{1}{2}\left[\frac{B'}{B} + \frac{A'}{A}\right]\dot{r}^2. \tag{5.35}$$

All we have to do is to apply $AB = 1$, and all vestiges of gravitational *repulsion* disappear.

Since we know that the geodesic equation is given by (2.78) it was unthinkable that the equation of the trajectory should give rise to the possibility of gravitational repulsion. As a check we can derive (2.78) directly from (5.35).

Introducing the values of the coefficients A, and, consequently, B, into (5.35) results in:

$$\ddot{r} - r\dot{\varphi}^2 = -g\left(V^2 + 3(\dot{r}\dot{\varphi})^2\right). \tag{5.36}$$

From the second geodesic equation, we know that the Schwarzschild metric conserves angular momentum, so introducing this condition into (5.36) gives:

$$\ddot{r} - r\dot{\varphi}^2 = -g\left(V^2 + 3\frac{L^2}{r^2}\right). \tag{5.37}$$

---

[11] McGruder, *op. cit.*
[12] S Weinberg, *Gravitation & Cosmology* (Wiley, New York, 1972).

Finally, transforming to $u/1/r$ results in

$$\frac{d^2u}{d\varphi^2} + u = \frac{\mu}{L^2}V^2 + 3\mu u^2,$$

which is precisely, Einstein's modification of the equation of the elliptical orbit, (2.78), with the additional parameter V introduced:

$$\frac{d^2u}{d\varphi^2} + u = \frac{\mu}{L^2}V^2 + 3\mu u^2. \tag{5.38}$$

5.5.2 Closed Orbits in the Schwarzschild field

Recall that the condition for constant velocity in the Schwarzschild field is:

$$V^2 = B\dot{t}^2 - A\dot{r}^2 - r^2\dot{\varphi}^2$$

in the plane $\vartheta = \pi/2$. Using the fact that $AB = 1$, and $B\dot{t} = 1$ is a first-integral, we rearrange the expression, and find for the conservative field, the energy integral:

$$\dot{r}^2 + r^2\dot{\varphi}^2 - \frac{\alpha}{r}V^2 - \alpha r\dot{\varphi}^2 = 1 - V^2, \tag{5.39}$$

where we introduced $B = 1 - \alpha/r$.

When we introduce the eikonal, S, according to $\dot{r} = \partial S/\partial r$, and $\dot{\varphi} = \partial S/\partial \varphi$, (4.3) becomes the Hamilton-Jacobi equation:

$$\left(\frac{\partial S}{\partial r}\right)^2 + r^2\left(1 - \frac{\alpha}{r}\right)\left(\frac{\partial S}{\partial \varphi}\right)^2 = \frac{\alpha}{r}V^2 + (1 - V^2). \tag{5.40}$$

This clearly testifies to the fact that we are in the realm of geometrical optics.

If we wish to differentiate (5.39), we change the differentiation variable to φ using the second integral of the motion, $r^2\dot{\varphi} = L$, and changing the dependent variable to $u = 1/r$, we again come out with Einstein's modification of the equation of the orbit, (5.38).

In the context of Gerber's theory of light deflection, the first term on the right-hand side of (5.38) must be made to vanish. In his book on the perihelion of Mercury, Roseveare [13] argues that

---

[13] N T Roseveare, *Mercury's Perihelion from Le Verrier to Einstein* (Oxford U P, New York, 1982).

to calculate the deflection of light we consider a particle approaching from infinity. For this we put $L = -\infty$.

Of course, this will get rid of the unwanted term, but a constant is a constant and that shouldn't change to accommodate the situation. Brown [14] argues that $L = r^2 d\varphi/d\tau$, where τ is proper time, and $d\tau$ vanishes along a light ray. But that's not how we defined the first integral, L, which is valid for any affine parameter just as long as its increment doesn't vanish.

The desired term can be eliminated quite easily by setting $V = 0$, which holds for light trajectories. Obviously, we must have $V < 1$, since we have set $c = 1$ in our discussion. This will have definite repercussions on the nature of the permissible trajectories.

But, the point to be made here lies with the energy integral (5.39). The differential equation for the orbit is:

$$\left(\frac{du}{d\varphi}\right)^2 = L^{-2}(1 - V^2) + L^{-2}V^2\alpha u + (1 - \alpha u)u^2.$$

Since the last term is a small perturbation, we can neglect it when determining the nature of the orbit:

$$\left(\frac{du}{d\varphi}\right)^2 = A + Bu - u^2,$$

where $A = (1 - V^2)/L^2$, and $B = \alpha V^2/L^2$.

The maximum and minimum values of u, corresponding to the perihelion and aphelion of the orbit are obtained as the roots of the quadratic equation:

$$u^2 - Bu - A = 0.$$

In order that there be two real roots $A < 0$, implying that $V^2 > 1$, which is physically impossible. This conclusion would have been impossible to reach had we considered only the second-order differential equation (5.38). This shows that the geodesics are of limited value.

---

[14] *op. cit.* §"Gerber's light deflection".

5.6 A miraculous cancellation?

The title probably comes from a remark made by Eddington, who upon dismissal of Laplace's determination of the speed of gravity, said:

> as we now know, the first-order which Laplace anticipated is compensated; the loss of energy is actually a residual effect of third-order of small quantities as determined by modern theory.

The small quantities Eddington is referring to are the relative velocities, and first-and third-orders pertain to the order of aberration.

The question is in which direction does the force of gravity point? If gravity propagated at the speed of light, it would create disastrous consequences for planetary orbits. The problem was posed by Eddington himself: [15]

> If the Sun attracts Jupiter towards its present position S, and Jupiter attracts the Sun toward its present position J, the two forces are in the same line and balance. But if the Sun attracts Jupiter to its previous position S', and Jupiter attracts the Sun toward its previous position J', when the force of attraction started out to cross the gulf, then the two forces give a couple. This couple will tend to increase the angular momentum of the system, and, acting cumulatively, will soon cause an appreciable change of period, disagreeing with observation if the speed is at all comparable with that of light.

Eddington gives a picture of this in Fig (5.3). Eddington is quick to add that the argument is 'fallacious', and should not be taken seriously. He gives the example of two electric charges that will point their forces in the direction of the present position, and not the past position even though "the electric influence is being propagated with

---

[15] A S Eddington, *Space Time and Gravitation* (Cambridge UP, Cambridge, 1929) p 94.

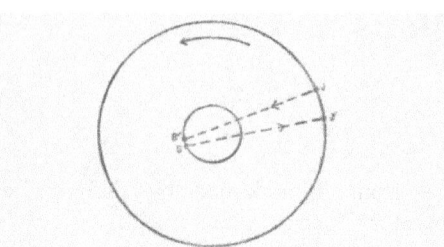

Figure 5.3: The retardation of the gravitational force creates a couple and renders the orbit of Jupiter unstable. Taken from Eddington first edition of Space Time and Gravitation

the velocity of light." He supports his argument by considering the retarded Lienard-Wiechert potential:

$$\phi = \frac{e}{[r(1 - v/c)]},$$

where v is the speed of the charge in the direction $\vec{r}$, and the bracket means to calculate the quantity at the past time. To first order in aberration, v/c, the denominator is equal to its "*present* distance r, so that the expression reduces to e/r in spite of the time of propagation."

More modern-day authors have called this a 'miraculous' cancellation of aberration terms at least to third-order so that a charge "in uniform motion... points toward the 'instantaneous' position so the effects of aberration are exactly canceled." [16]

Carlip is not impervious to the problems that beset GR. He first mentions correctly that "there is no preferred time-slicing in general relativity, and thus no unique definition of an 'instantaneous' direction." If a flat Minkowski background is used, he expects "ambiguities" to second order in the relative velocity.

Second, Carlip claims that:

> we cannot simply require by fiat that a massive source accelerate. The
> Einstein field equations are consistent only when all gravitational sources

---

[16]S Carlip, "Aberration and the speed of gravity," arXiv:gr-qc/9909087v2.

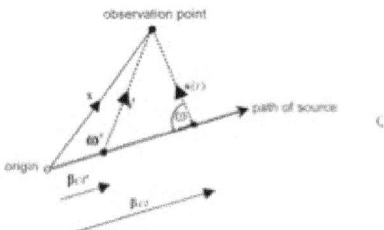

Figure 5.4: A uniformly moving charge as seen by a distant observer. Taken from IPL *op. cit.*

move along the trajectories determined by their equations of motion, and, in particular, we can consistently represent an accelerated source only if we include the energy responsible for its acceleration.

Carlip first claims that Einstein field equations make particles follow geodesics, but that the equations of motion allow for accelerations—if we include the energy for their acceleration. Where do we include this energy if energy cannot be localized in GR? Moreover, unlike Maxwell's field theory where the equations of motion (Lorentz force) are independent of the field equations, GR has the field equations determining the motion, and that motion is uniform, at constant velocity.

In another attempt to show that there is a miraculous cancellation of aberrational terms in the field, Ibison *et. al.* [17] argue on the basis of analogy between EM and GR that since an electric field of a uniformly moving charge, composed entirely of retarded influences, "will *accelerate* a remote test charge towards the instantaneous position of the source." Hence, a similar result can be expected in GR.

Specifically, IPL consider a charge in uniform motion, as shown in Fig (5.4) Let $\vec{s}(t)$ denote the vector from the current position of the charge to the observer, and $\vec{x}$ be the directed distance from the origin to the observer so that

$$\vec{s} = x - ct\vec{\beta} = \vec{s}' - \vec{\beta}s', \tag{5.41}$$

where the last equality expresses it in terms of the past position $\vec{s}'$, and because the motion is uniform, $\beta = \beta'$.

---

[17] M Ibison, H E Puthoff & S R Little (IPL), "The speed of gravity revisited".

Considering only the gradient of the electric field contribution to the electric field, (cf. Eq (5.44) below), they express the electric field,

$$\vec{E} = e\frac{\gamma^4 \vec{s}(t)}{\sigma^3}, \tag{5.42}$$

in terms of the hyperbolic distance,

$$\sigma = \frac{s(t)\sqrt{1 - \beta^2 \sin^2 \omega}}{1 - \beta^2} = \frac{\sqrt{(\vec{x} - ct\vec{\beta})^2 - (\vec{x} \wedge \vec{\beta})^2}}{1 - \beta^2} \equiv \gamma^2 s K(\omega), \tag{5.43}$$

where ω is the angle formed from $\vec{\beta}$ and $\vec{s}$, at the present position, $\gamma = 1/\sqrt{1 - \beta^2}$ is the Lorentz contraction factor, and

$$K(\omega) = \sqrt{1 - \beta^2 \sin^2 \omega}.$$

From (5.42) IPL conclude that "the electric field of a uniformly moving particle is not aberrated." The transformation to the hyperbolic distance caused by the retardation of the field, (5.43), which they don't recognize as such, only "affects the magnitude, but not the direction, of the field."

So the retardation of the propagation of the electric field transforms Euclidean into hyperbolic, or Lobachevski, geometry, and this should lead to aberration. For consider two infinitesimal elements of length in Lobachevski space:

$$ds^2 = \frac{c^2 (d\vec{v})^2 - [\vec{v} \wedge \vec{v}]^2}{(c^2 - v^2)^2},$$

for neighboring velocities $\vec{v}$ and $\vec{v} + d\vec{v}$, and

$$\delta s^2 = \frac{c^2 (\delta \vec{v})^2 - [\vec{v} \wedge \delta \vec{v}]^2}{(c^2 - v^2)^2}$$

for neighboring velocities $\vec{v}$ and $\vec{v} + \delta \vec{v}$. The cosine of the angle between the displacements is:

$$d\vec{s} \cdot \delta \vec{s} = ds\delta s \cos \alpha = \frac{c^2 (d\vec{v} \cdot \delta \vec{v})^2 - [\vec{v} \wedge d\vec{v}] \cdot [\vec{v} \wedge \delta \vec{v}]}{(c^2 - v^2)^2}.$$

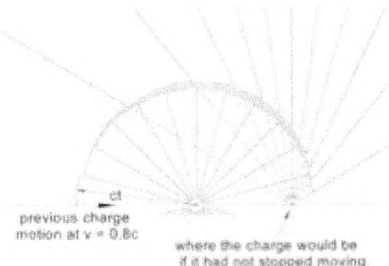

previous charge
motion at v = 0.8c

where the charge would be
if it had not stopped moving

Figure 5.5: The curving of the field lines when a charge is suddenly brought to rest. Taken from IPL *op. cit.*

This can be considered as the definition of an angle in Lobachevski space, and it is an angle of aberration. [18] Velocity vectors for a triangle in Lobachevski space whose sum is less than two right angles by the magnitude of the aberration. Thus, Lobachevski space has aberration incorporated into it!

Apart from the fact that a moving force is ambiguous, and to which space it refers, the electric field on a charge at any given instant in time is due to the field at that instant. A uniformly moving charge in that field is subjected to a force that is proportional to and directed in the same direction as the electric field at its present position so the force *always points directly toward the source at all times.*

In an inertial fame moving with the charge, the charge will not feel a magnetic field; thus the force is entirely made up of the electric field and it will always point directly to the source at all times.

IPL admit that when the charge is accelerated that their miraculous cancellation of terms in the electric field no longer is valid. The electric field thus acquires terms that depend on the acceleration so that "the charge is not, in general, directed toward the accelerating source." They illustrate this in Fig (5.5) when an initially uniform charge is suddenly brought to rest.

---

[18]V Fock, *The Theory of Space Time & Gravitation* (Pergamon, London, 1959) pp 43-44.

To explain how such a phenomenon occurs, and why it will involve terms of order β^2 or higher, they introduce the total electric field:

$$\vec{E} = -\nabla\phi - \frac{1}{c}\frac{\partial \vec{A}}{\partial t}, \tag{5.44}$$

where $\vec{A}$ is the vector potential of the EM field. Writing the components of as a 4-vector as Lienard-Wiechert potentials:

$$\{\phi, \vec{A}\} = \frac{e\gamma^2\{1, \vec{\beta}\}}{\sigma},$$

IPL conclude that "it is readily deduced that the contribution to the electric field from the *time derivative* of the vector potential is of order β^2 or higher and cannot therefore play a role at low velocities."

What IPL are referring to is contraction in the direction of the motion. The diagram Fig (5.5) can be interpreted as the contraction of equi-potential surfaces of a moving charge in the direction of the motion. Instead of the radial field lines being everywhere normal to spheres, it appears that the sphere becomes an ellipsoid, and the gradient of the scalar potential does not point toward the center where the source is located.

Both scalar and vector potentials satisfy inhomogeneous wave equations:

$$\nabla^2\phi - \frac{1}{c^2}\frac{\partial^2\phi}{\partial t^2} = -4\pi\rho, \qquad \nabla^2\vec{A} - \frac{1}{c^2}\frac{\partial^2\vec{A}}{\partial t^2} = 4\pi\rho\frac{\vec{v}}{c}, \tag{5.45}$$

where ρ is the charge density. The two wave equations are interchangeable when we write:

$$\vec{A} = \vec{\beta}\phi. \tag{5.46}$$

This clearly shows that the vector potential is a first-order aberration of the scalar potential!

At constant velocity it follows that:

$$\frac{\partial\vec{A}}{\partial t} = \vec{\beta}\frac{\partial\phi}{\partial t}.$$

Since the total time derivative of the scalar potential, $d\phi/dt$, vanishes along geodesics because ϕ is stationary, it follows that:

$$\frac{d\phi}{dt} = \frac{\partial\phi}{\partial t} + \vec{v} \cdot \nabla\phi = 0.$$

Hence, the electric field becomes:

$$\vec{E} = -\nabla\phi + \vec{\beta}(\vec{\beta} \cdot \nabla\phi). \tag{5.47}$$

Choosing any direction as that of the motion of the charge, (5.47) will show second-order contraction in that direction. Or, equivalently, if surfaces of constant potential appear spherical for a charge at rest, they will transform into ellipsoids when in motion due to such contraction. This is essentially what Fig (5.5) illustrates.

But, we are talking about first-order field effects which (5.44) split up into when expressed as retarded potentials. For (5.44) naturally splits up into 'velocity fields,' which IPL considered, which are independent of acceleration, and "acceleration fields," which depend linearly on $\dot{\vec{\beta}}$: [19]

$$\vec{E}/e = \gamma^4 \frac{\vec{s}' - \vec{\beta}'s'}{\sigma^3} + \frac{\gamma^6}{c} \frac{\vec{s}' \wedge \left[(\vec{s}' - \vec{\beta}'s') \wedge \dot{\vec{\beta}}'\right]}{\sigma^3}, \tag{5.48}$$

where σ is given by: (5.43), $\beta' \neq \beta$, since the motion is no longer uniform, and it is calculated at the past position.

Moreover, the second term in (5.48) has a component pointing in the direction of the accelerating charge:

$$\vec{E}_{rad} = \frac{1}{c} \frac{\vec{s}(t)(\vec{s}' \cdot \dot{\vec{\beta}}') - \dot{\vec{\beta}}'(\vec{s}' \cdot \vec{s})}{(s' - \vec{\beta}' \cdot \vec{s}')^3}, \tag{5.49}$$

which, in general, does not coincide with the instantaneous direction to the source.

Furthermore, it is only the second, or radiative, term in (5.48) that contributes to the Poynting vector,

$$\vec{S} = \frac{c}{4\pi} \vec{E}_{rad} \wedge \vec{B}_{rad}, \tag{5.50}$$

---

[19] J D Jackson, *Classical Electrodynamics* (Wiley, New York, 1975) §14.2.

where $\vec{B}_{rad}$ is the radiation component of the magnetic field,

$$\vec{B}_{rad} = \frac{1}{c^2} \frac{\vec{s}' \wedge \dot{\vec{\beta}}'}{(s' - \vec{\beta}' \cdot \vec{s}')^2} \tag{5.51}$$

and, consequently, to the power radiated by the accelerating charge,

$$\frac{dP}{d\Omega} = \frac{e^2}{c} \left| \vec{n}' \wedge \left(\vec{n}' \wedge \dot{\vec{\beta}}' \right) \right|^2 , \tag{5.52}$$

in a unit solid angle, where $\vec{n}' = \vec{s}'/s'$ is a unit vector calculated at the past position. Both the Poynting vector (5.50) and the power radiated in a unit solid angle, (5.52), are beyond the limits of GR, because GR constrains particles to follow geodesic paths. The pseudo-tensor, (6.24) below, cannot fill the void.

Hence, both IPL and Carlip were looking in the wrong place entirely. Why should Carlip have looked to the connection, Γ^i_{00}, for a particle initially at rest, to determine "the 'acceleration' of such a particle"? Because GR prevents the inclusion of acceleration terms, like that in (5.48), due to the fact that particles are constrained to follow geodesic paths. Thus, the connection then is the only thing that remains to give the culprit to. However, this is a grave defect in GR, and not in the physics of the problem.

5.7 Do forces aberrate?

Newton says they do. Returning to our discussion of Fig (2.1), the centripetal force F_c is directed toward the origin O, while the force F_0 is directed to the source S at center of the osculating circle. The ratio of the two, F_c/F_0 is related just as the ratio of the angular momentum to the linear momentum L/rv are by $\sin \alpha$, which for small enough α is proportional to α itself, the angle of aberration. However, when the two are combined, under the condition that angular momentum is conserved, the aberration becomes third-order.

Since Newton replaces the motion along an incremental arc of a general curve at point P with *uniform* circular motion along an incremental arc of a circle of radius ϱ at

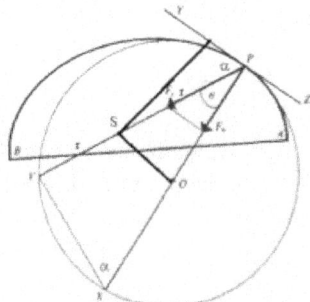

Figure 5.6: A modification of the revised diagram for Newton's basic dynamic theorem where the source divides the chord PV in half leading to uniform motion and first-order aberration.

the same point, we should use constant angular velocity and not constant angular momentum. In that case the source S should find itself at the center of the chord PV of the osculating circle, as shown in Fig (5.6).

Now, the 'circular dynamics ratio' $PV/2PS = \varrho \sin \alpha / r$ implies:

$$PV/2 = \varrho \sin \alpha = 2\frac{1 + u'^2/u^2}{u + u''},$$

in terms of the variable $u = 1/r$, where the primes stand for differentiation with respect to ϑ. The right-hand side must give r, which is satisfied by a circle, $r = R \sin \varphi$ of arbitrary radius, R. This is uniform circular motion with $r/\varrho = \sin \alpha$. The $\triangle SPO$ becomes a right-triangle with $\cos \vartheta = \varrho/r$.

But, Newton is not considering uniform circular motion for Kepler's II, (2.2), must be satisfied. So Newton turns his attention to the triangle, $\triangle VPS$, formed from linearly extrapolating the position at P so that the new position of the body will feel the force of the source head-on. Now the circular dynamics ratio is:

$$\frac{SY^2}{SP^2} = 2\varphi \sin^3 \alpha = 2\frac{L^2}{u + u''} = \text{const.}$$

where the constant contains the product $F_0 r^2$. This says that the curvature is due to a third-order aberration, in contrast to the first-order aberration in the case of uniform circular motion. So, when the centripetal and central forces do not point in the same direction we can rightly expect aberration to occur, like the angle that rain-drops hit the windshield of a car in uniform motion.

True, Newton considered the propagation of gravity as instantaneous. But, here he is considering two positions of the projectile, one a P and another some 'time' later to be at the linearly extrapolated position Y, as if it could move rectilinearly. The difference between its true position on the curve gives rise to the angle of aberration, as shown in Fig (5.1).

Let us take for granted, for the moment, that Snell's law holds on geodesics. [20] Applying the law of sines to Fig (5.4) we get:

$$\frac{\sin(\omega' - \omega)}{\beta s'} = \frac{\sin \omega}{s'} = \frac{\sin \omega'}{s}.$$

The relative indices of refraction can be read off from the equalities. In the first equality the relative index of refraction is just what we would expect, $n = 1/\beta = c/v$, while the second equalities gives $n = s/s'$. These show that Snell's law is obeyed, and that we are dealing with geodesics, or paths of constant velocity.

By computing $s'^2 = (\vec{s} + \vec{\beta}s')^2$, we find:

$$s'/s = \frac{\beta \cos \omega + K(\omega)}{1 - \beta^2}. \tag{5.53}$$

And introducing this into

$$s' - \vec{s} \cdot \vec{\beta} = (1 - \beta^2)s' - s\beta \cos \omega,$$

we come out with:

$$s'/s = \frac{K(\omega)}{1 - \beta \cos \omega'}. \tag{5.54}$$

Both (5.53) and (5.54) were derived by IPL. However, what they failed to do was to equate them, for then they would have obtained:

$$\cos \omega = K(\omega)\frac{\cos \omega' - \beta}{1 - \beta \cos \omega'}. \tag{5.55}$$

---

[20]We will show this to be true in the next Chapter in the section dealing with geodesics.

For $K(\omega) = 1$, implying that $\omega = 0, \pi$, (5.55) reduces to the usual expression for aberration. [21]

In fact, the relativists use a special form of the aberration formula (5.55) when they set $\omega = \pi/2$ so that $\beta = \cos\omega$. For then either (5.53) or (5.54) reduce to a space contraction: $s = s'\sqrt{1 - \beta^2}$, or a second-order Doppler shift that was experimentally verified by Ives and Stilwell in 1938. [22]

Where aberration should manifest itself, and does, is in the tangential force, and not a force that is directed along the motion.

5.8 Can the universe be collapsed into a nutshell?

Consider the metric used in the Kennedy-Thorndike (KT) inferometer:

$$c^2(1 - \beta^2)dt^2 = ds^2 \pm 2v \cos\omega dt ds, \tag{5.56}$$

where the angle ω is formed from the arm of the interferometer and the velocity of the 'ether', $\vec{v}$. The $\pm$ refer to the forward and backward journeys of the split beam of light.

By completing the square in the KT-metric (5.56) and taking the positive square-root, we obtain ratio of velocities as:

$$c/ds/dt = \frac{\pm\cos\omega + K(\omega)}{1 - \beta^2}, \tag{5.57}$$

where

$$K(\omega) = \sqrt{1 - \beta^2 \sin^2\omega}. \tag{5.58}$$

---

[21] cf. B H Lavenda, *A New Perspective on Relativity* (World Scientific, Singapore, 2011) Eq (8.39).

[22] H E Ives & G R Stilwell, "An experimental study of the rate of a moving atomic clock," *JOSA* **28** (1938) 215-226.

The ratio, (5.57), conforms to the ratio of the retarded and present distances to the observer, (5.53), that we derived in the last section, only now the relative velocity β can be of both signs,

$$s'/s = \frac{\pm \beta \cos \omega + K(\omega)}{1 - \beta^2}.$$

Since $s' = c(t - t')$, the time lapse to go from past to present position, we may associate s' with cdt, and s to its increment, ds.

The ratio of the distance from past and present distances to the observer is also given by (5.54). Hence, we may equate the two expressions to obtain:

$$\cos \omega' = \frac{\beta \pm \cos \omega / K(\omega)}{1 \pm \beta \cos \omega' / K(\omega)}, \tag{5.59}$$

which would be the usual formula for aberration $\omega = 0, \pi$.

As we saw in the last section, the factor $K(\omega)$ figures into the definition of hyperbolic distance:

$$\frac{s' - \vec{\beta} \cdot \vec{s'}}{1 - \beta^2} = \frac{sK(\omega)}{1 - \beta^2} = \frac{\sqrt{s^2 - (\vec{\beta} \wedge \vec{x})^2}}{1 - \beta^2} = \frac{\sqrt{(\vec{x} - ct\vec{\beta})^2 - (\vec{\beta} \wedge \vec{x})^2}}{1 - \beta^2}, \tag{5.60}$$

where we recall that $\vec{x}$ is the vector from the origin to the observation point in Fig (5.4).

In fact, the factor (5.58) is related to same factor at the past position by:

$$K(\omega') = \frac{K(\omega)}{1 - \beta^2}.$$

This results in a 'hybrid' relation:

$$\cos \omega' = \frac{\beta \pm \cos \omega}{1 - \beta^2},$$

between aberration and Doppler.

What prevents (5.58) from being unity is the tilt of the tilt of the chord AB with respect to the diameter of the circle in Fig (5.7). The perpendicular distance from the center of the circle O to the chord is $v \sin \omega$.

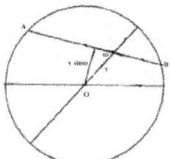

Figure 5.7: The tilt of the chord AB with respect to the diameter of the circle prevents K from being unity. The angle ω can also be interpreted as the inclination of two planes with respect to one another.

Instead of the KT-metric, (5.56) we can consider space contraction:

$$c^2 dt^2 = (1 - \beta^2)ds^2 \pm 2b\cos\omega\,dt\,ds. \tag{5.61}$$

Completing the square and taking the positive square root give:

$$ds/dt/c = \frac{\pm\cos\omega + K(\omega)}{1 - \beta^2}, \tag{5.62}$$

which is precisely the inverse of (5.57).

All interferometers measure the two-way speed of light so that on the completion of the round-trip the linear terms in the velocity, $\pm\beta\cos\omega$, cancel out. Consequently, interferometers are impervious to aberration since these terms cancel in the final summation.

Denote,

$$c'' = c\frac{1 - \beta^2}{K(\omega)} = \frac{c}{K(\omega)'} \tag{5.63}$$

the KT-speed of light, which can be considered as an index of refraction due to the motion. Analogously, call

$$c' = c\frac{K(\omega)}{1 - \beta^2} = cK(\omega'), \tag{5.64}$$

the Beltrami-Lobachevski (BL) speed of light. Both (5.63) and (5.64) are two-way speeds.

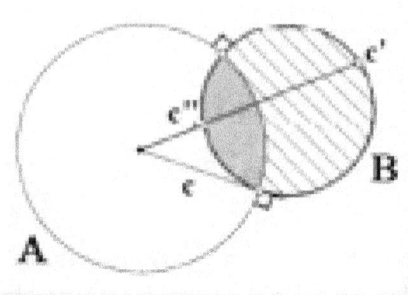

Figure 5.8: The diagram illustrates inversion where every circle orthogonal to a given circle gets mapped into itself. The circle B gets 'mapped into itself' meaning that the shade and hatched regions into which the circle A divides it is being swapped under inversion.

Now, the beauty of this exercise is that their product, $c''c' = c^2$, like the product of phase and group velocities. That is, they are inverses of one another with respect to an invariant circle of radius c, as shown in Fig (5.8). That both speeds lie on one side of the origin of A, and c' lies outside circle A mean that we are dealing with hyperbolic geometry . It is for this reason that we called (5.64) the BL two-way speed of light. On the strength of inversion, if the KT speed exists so, too, must the BL speed.

Inversion is rather miraculous: it can turn squares into circles. Fig (5.9)

shows a circle of radius c about the origin, O, of a chess board. By the property of inversion, every point P that lies outside the circle is paired with a point inside the circle, P' such that $\overline{OP'} \cdot \overline{OP} = c^2$. Lines are also paired; a line extending outward toward the infinite corresponds to one pointed inward toward the center, but no matter how far the outward line extends, the inward line will never reach the center.

Martin Gardner tells the mathematical joke about how to catch a lion: build a cage and perform inversion. The universe itself can be inverted and compressed within a golf ball.

Figure 5.9: On account of inversion every point on the chess board outside the circle i put into one-to-one correspondence with every point inside the circle. Lines extending outward to infinity are paired with lines inside the central white space, that are directed inward to the center but never attaining it. Figure taken from M. Gardner, *6th Book of Mathematical Discoveries from Sci Am* (U Chicago Press, Chicago, 1984) p 245.

However, instead of considering distances we have considered speeds, and, in particular the speed of light. And, there is nothing 'sacred' about the invariant circle.

If the velocity of light were a vector, depending on orientation, the radius of the circle would be $\vec{c} \pm \vec{v}$. The length of a chord would be the square-root of the difference between the square of the radius, $\vec{c} - \vec{v}$ and the square of the perpendicular distance from the chord to the center of the circles, as shown in Fig (5.7),

$$cK(\omega) = \sqrt{(\vec{c} - \vec{v})^2 - (\vec{c} \wedge \vec{v})^2} = \sqrt{(\vec{c} - \vec{v})^2 - v^2 \sin^2 \omega}, \qquad (5.65)$$

or, equivalently,

$$cK(\omega) = c - v \cos \omega.$$

This is a *one-way* speed.

The expression corresponds to the one:

$$c(\omega') = 1 - v(1 + 2\alpha) \cos \omega', \qquad (5.66)$$

reported by Mansouri-Sexl (MS), [23] which is a directional quantity for $\alpha \neq 1/2$. Observe that the angle is at the retarded position, and the constant α is related to the quadratic term in $t' = at + \vec{v} \cdot \vec{x}$, where $a = 1 + \alpha v^2 + \cdots$. In fact, it is zero.

In the case of Einstein synchronization, MS found: [24]

$$v(\omega) = \frac{b(1 - \beta^2)}{a\left[\cos^2 \omega + b^2 d^2 (1 - \beta^2) \sin^2 \omega\right]^{1/2}}.$$

This is a second-order speed of light, and does not depend on the direction in which the light ray moves. There is no first-order effect. The coefficients a, b and d were considered as arbitrary functions of β, that appear in the transform:

$$t' = at + vx, \qquad x' = b(x - vt), \qquad y' = dy, \qquad x' = dz.$$

As a matter of fact, if the MS expression is to coincide with the KT-speed, c'', then $a = b = d = 1$.

5.9 What is the speed of light, anyway?

Never have two numbers been so similar and so far apart. Maxwell proposed that the speed of light c, be identical to his constant, b. The latter he found as the ratio of two completely, and entirely, different forces: the Coulomb, F_E, and magnetic, F_M, forces. For the latter, he used the Biot-Savart law .

Einstein in 1905 took this to the next level by claiming that, in all inertial systems, "the speed of a ray of light in vacuum is constant, independent of the moving body." This is essentially his second postulate of his 'special' theory of relativity, and it muddied the waters still further.

In doing so, he negated the Doppler effect , as Essen [25] emphasized. For if c is the ultimate speed, as Einstein argued,

---

[23] R Mansouri & R U Sexl, "A test theory of special relativity: I. Simultaneity and clock synchronization," *Gen Rel & Grav* **8** (1977) 491-513.

[24] *op. cit.* Eq (6.1.7).

[25] L Essen, *The Special Theory of Relativity: A Critical Analysis* (Clarendon Press, Oxford, 1971).

> instead of obtaining the values $c + v$ or $c - v$ for the velocity of light,
> where v is his is own velocity relative to the source, an observer obtains
> the values [*sic*] c. Thus it appears that there should be no Doppler change
> in frequency, and yet this effect is known to exist.

And if c is an absolute constant, isn't still equal to the ratio of the distance to the time
that it takes to cover that distance, namely $c = d/t$?

> It is possible to define units of any two of the quantities in this expres-
> sion ... making the velocity of light have the constant value c even to
> observers in relative motion is comparable to making it a unit of measure-
> ment.

Nevertheless, experts in time measurement like Essen, the inventor of the cesium clock,
and Ives, who verified the second-order Doppler effect, $\sqrt{1 - v^2/c^2}$, predicted by Einstein, failed to make the distinction between electromagnetic and kinematic definitions
of the speed of light. The former definition was, in fact, a mere 'ratio of units,' while
the latter was the ratio of the distance light travels in a given time interval.

Both Essen and Ives were anti-relativists. Essen considered the theory of relativity
"simply as a new system of units it can be made consistent but it serves no useful purpose." Ives, on the other hand, thought that his experiment verified that moving clocks
slow down and moving rods contract, but had nothing to do with special relativity
which he saw as inconsistent. [26]

The accepted value of the speed of light, $c = 299792458$ m/s, and it was determined by
"the length of the path traveled by light in vacuum during a time interval $1/29979458$
of a second." Essen argued that special relativity dismissed both the independence of
the space and time measurements, and, consequently wrote off special relativity as a
"mere change in units."

Does the electromagnetic field exist *per se* independent of ponderable matter? There
is no way of knowing this since all experiments that have been devised require it to

---

[26] *The Einstein Myth and the Ives Papers* R Hazelett & D Turner eds (Devin-Adair, Greenwich CT, 1979) p 58.

interact with matter. Nonetheless, Maxwell constructed his electromagnetic theory independently of matter, using instead his 'ether' which was supposedly the medium that allowed for the propagation of electromagnetic waves.

Again, by decree, Einstein abolished the ether, which he resurrected over a decade later in his GR. For he considered GWs to propagate as 'ripples' in spacetime making them contract and dilate successively.

If there is no ether, then what do the constants ϵ_0 and μ_0, the electric permittivity and magnetic permeability, refer to? And why should their product turn out to be a universal constant far away from the reaches of ponderable matter?

These constants appear as constants of proportionality in the ratio of the magnetic to Coulomb forces. Coulomb defines the repulsive electrostatic force, F_E, as the action of a point charge, e, on another like charge, e', that are separated by a distance r. Both charges are assumed to be at rest with respect to one another so that Coulomb's force reads:

$$F_E = -\frac{ee'}{4\pi\epsilon_0 r^2}.$$

The constant of proportionality, $4\pi\epsilon_0$, is arbitrary insofar as it refers to the units that are chosen.

It was known since the dawn of the nineteenth century, that current elements ids and $i'ds'$ in infinitesimal conductors of length ds and ds' repel each other also through an inverse-square law:

$$F_M = \frac{\mu_0}{4\pi}\frac{ii'dsds'}{r^2},$$

just like Coulomb's law, where the constant of proportionality $4\pi\mu_0$ also depends on the units that are chosen.

Yet, these two constants of proportionality can't be independent of one another since currents consists of charges. In order to unite the static Coulomb law with the kinetic Ampère law, Maxwell found it necessary to introduce a constant, b in:

$$bids = ve,$$

where v is the velocity of the moving charge e. So said Maxwell [27] b would represent "the number of units of statical electricity which are transmitted by a unit electric charge in a unit of time." Actually, there is no need of such constant since it will emerge spontaneously from the ratio of the magnetic to electric forces.

The ratio of magnetic to electric forces is:

$$F_M/F_E = -\epsilon_0\mu_0 vv',$$

where we have set $ev = ids$, and $e'v'i'ds'$. This immediately suggests that F_M is merely an electric force in motion, and, consequently, should not be considered as a separate entity. The ratio also suggest that the motional electric force is a second-order quantity in the velocity of the charges.

Since the ratio of the forces is dimensionless, the quantity:

$$b = 1/\sqrt{(\epsilon_0\mu_0)},$$

must also have dimensions of a speed. b is often referred to as Maxwell's constant. If, as Maxwell supposed, that b is c, then only one of the two quantities, ϵ_0 and μ_0, can be independent. But, this avoids completely static measurements of these constants which are properties of ponderable matter.

Quite astonishingly, the last static measurements of b were made by Rosa and Dorsey (by coincidence) in 1905 — the year heralded as Einstein's miraculous year. Of the three 'ground-breaking' papers published that year, one was dedicated in rendering superfluous the very difficult electrical measurements of Maxwell's constant, b.

The static measurements performed by Rosa and Dorsey, performed in 1905 but published two years later, found a significant difference between Maxwell's::

$$b = 29972000 \pm 20000 \text{ m/s},$$

and c found kinetically. For, in fact, the measurement of b had nothing to do with the measurement of the speed of light. In Maxwell's own words:

---

[27] J Clerk Maxwell, *A Treatise on Electricity & Magnetism*, vol. II (Clarendon Press, Oxford, 1873). §849.

The value of b was determined by measuring the electromotive force which a condenser of known capacity was charged, and then discharging the condenser through a galvanometer, so as to measure the quantity of electricity in it in electromagnetic measure.

This echoes Maxwell's definition of c as the number of *statical* units given above. No use, whatsoever, was made of light in the process of measurement.

In contrast, the value of c found by Foucault was derived from measurements of the angle through which a revolving mirror rotated while the light reflected from it went and came back along a predetermined course. Neither charges nor currents entered into the determination of c. However, unlike the kinematical measurements of c, the static measurements of b were entirely lacking— thanks, of course, to Einstein.

Maxwell's equations in the absence of sources decree that the electromagnetic field is either longitudinal and stationary, or transverse and propagating at a speed b. Since sources are absent, it cannot be concluded, however, that the EM wave that was created by an oscillating electric dipole spreads outward from its source at speed b.

In the presence of ponderable matter, the speed at which energy propagates, as determined by the direction and magnitude of Poynting's vector, is appreciably smaller than b. Only for 'pure' electromagnetic waves are the intensities of the electric, E, and magnetic, B, fields related by $E^2 = b^2 B^2$. The proportionality factor, b^2, refers to the 'radiation' zone, which is deprived of ponderable matter.

Recalling our earlier discussion of which way the electric field points as it goes from its past to present position, depends on the direction of the acceleration vector, and this need not be in the same direction as the position vector, $\vec{s} = \vec{s'} - \vec{\beta}s'$, for a mobile moving a constant speed.

If the magnetic and electric forces are independent, their sum is:

$$F_{EM} = F_E + F_B = F_E \left(1 - \frac{v^2}{b^2}\right),$$

for a pair of charges traveling at a common speed, v. This testifies to the fact that the total force is decreased in the direction of the motion by a factor, $1 - v^2/b^2$. Strangely enough, this is off by a factor of one-half in the square of the ratio of speeds according to the Weber and Kohlrausch measurements. Nevertheless, it does point to the fact that Maxwell's b should represent an upper limit to all accessible speeds of the two charges.

In order to discriminate between the possible speeds of light, let us return to the KT-speed of light, (5.63). The quotient on the right-hand side of that equation will be appreciated as the increments in the ratio of euclidean and hyperbolic arc lengths.

Consider a simple Michelson interferometer of length, L, and let v be the speed of the ether. The total time to complete a back and forth journey is: [28]

$$T = \frac{L}{c + v} + \frac{L}{c - v} = \frac{2Lc}{c^2 - v^2}. \tag{5.67}$$

Defining the two-way speed of light as $c'' = 2L/T$, we obtain the expression:

$$c'' = \frac{c^2 - v^2}{c} = c_G^2/c_A = c_H, \tag{5.68}$$

where

$$c_G = \sqrt{(c + v)(c - v)},$$

the geometric mean, and

$$c_A = (1/2)(c + v + c - v),$$

the arithmetic mean, speeds. The speed c_H in (5.68) is the harmonic mean speed.

Consequently, the KT's speed of light, c'', is the harmonic mean speed, c_H. It is this speed, which lies inside the invariant velocity circle in Fig (5.8), and it is the smallest possible speed of light attainable.

If we were to agree with Einstein that c is the invariant, upper limit of all speeds how do you explain all these other speeds of light?

---

[28] Like the expression for the longitudinal Doppler effect, no one has ever questioned why the speed in the outward journey should be $c + v$, if c is the limiting speed of light. In French's book, *Special Relativity* (W W Norton, New York, 1966), it corresponds to the formula before Eq (2-7), not even being dignified with a number of its own.

When Michelson rotated his apparatus by 90 degrees, he found a new time difference, and the difference between the two orientations would definitely lead to a shift of the interference pattern. Yet, no such pattern was noticed. To patch things up, FitzGerald, and slightly later Lorentz, concocted an explanation where the arm of the interferometer in the direction of motion was contracted by a factor of a second-order Doppler shift.

Whereas the KT-metric (5.56) predicts a time dilation, the metric (5.61) predicts a space contraction, but with a speed of light given by (5.64). Thus, a greater speed of light is the same as if the speed of light were c, but with the arm contracted. The BL-speed: $c' = c_A^3/c_G^2 = c^2/c_H$, or the inverse of the KT-speed, (5.68), would be the greatest attainable speed, and lies outside the invariant circle in Fig (5.8).

All this must seem strange when we have had it drilled into us that the invariant speed, c, or the arithmetic average, is the largest attainable speed. The resolution of this seemingly paradoxical situation lies in the factor $cK(\omega)$ in Eq (5.65), for when it is divided by the factor $1 - \beta^2$ it becomes a distance in Lobachevski space, and not any more Euclidean space.

5.10 The speed of gravity revisited

In an elliptical orbit, let a_0 be the initial value of the semi-major axis at time t_0. At any other time perturbation theory has shown that the semi-major axis will evolve in time according to (5.2).

Its rate of change can be related to the angular speed, ω, through Kepler's III by taking it to be a constraint, $a^3\omega^2 = $ const. The rate of change of the period, $P = 2\pi/\omega$, will be given by third-order aberration: [29]

$$\dot{P} = 2\pi \left(\frac{a}{c} \frac{2\pi}{P} \right)^3, \tag{5.69}$$

where a/c is the time it takes light to travel a distance a. Is this the same as if the

---

[29]We have shown that if the orbit were circular, first-order aberration would apply; for elliptical orbits Newton showed that third-order aberration applies.

speed were v_G instead of c?

Setting the time derivative of (5.2) equal to (5.69) via Kepler's constraint, gives:

$$\frac{v_G}{c} = 3 \left(\frac{c}{a\omega} \right)^2. \tag{5.70}$$

Now, SR demands that the angular speed, $a\omega < c$, so the phase velocity of the GW, $v_G > c$.

If gravity, or GWs, behaved like light, their finite speed of propagation would imply that the force acquire a small tangential component when it acts on a moving body. If this were first-order aberration, the force acting on the Earth would v_E/c or 10^{-4} that of the radial gravitational force, where v_E is the speed of the Earth that is 10^{-4} times less than the speed of light.

This tangential force, rather than slowing the Earth down, like an electromagnetic radiation pressure would, would tend to speed up the Earth. Since such an effect would be cumulative, it would double the Earth's distance from the Sun every 1200 years.

Since this does not happen, we can safely conclude that the Sun does not act from a retarded position with respect to the Earth, as it would do with Jupiter in Eddington's Fig (5.3). We saw in the first section of this chapter how Laplace reasoned from how long it would take the Moon to collide with the Earth, based on Halley's assumption, and given the Moon's lifetime at some 30 billion years found the speed of propagation to be 7 million times faster than the speed of light.

6 The electrodynamics of GWs

In this chapter, we will investigate the relation between GWs and EM waves where the average energy, instead of being given by Planck's law, is given by Newton's. Application of Kepler's III transforms the luminosity into a 7th-order, rather than a 3rd-order black-body aberration law.

6.1 Sticky beads, 'sticky' arguments

In the 1950's, when GWs were hotly debated, Richard Feynman tried to make the existence of gravitational waves (GW) palatable through an analogy with sticky beads on a rigid stick which is placed perpendicularly with respect to a passing GW, as shown in Fig (6.1). Feynman's purpose, as he openly admitted in a personal letter, was "to convince people that gravitational waves must carry energy." The question is what makes the beads move along the sticky stick?

According to Feynman, GW generate tidal forces about the midpoint of the stick. Since the stick is held transversely with respect to a passing GW, they can only produce longitudinal expansion and rarefaction oscillations, much like sound waves, in sliding the beads along the stick, as shown in Fig (6.2). Because the contact between the beads and stick are 'sticky,' some of the gravitational energy will be transformed into 'heat'. The fact that GWs produce heat is a clear sign that they carry both energy and momentum, and the work they do winds up as dissipated energy in the form of heat.

The argument hinges on *energy conservation*: what is lost as useful work shows up as dissipated heat. The argument can be made more quantitative, if not more transparent,

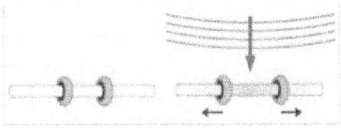

Figure 6.1: The "sticky" beads rationalization of a passing GW.

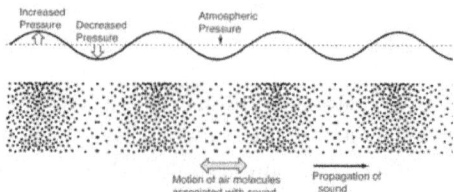

Figure 6.2: Sound waves are propagated longitudinally by alternating compressions and rarefactions.

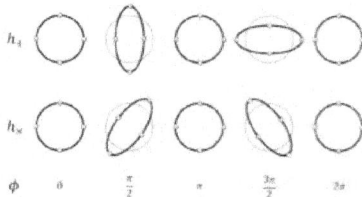

Figure 6.3: The passage of a linearly polarized GW through a ring of particles in the orthogonal direction to the plane. The circles shrink in one direction and expand in the orthogonal direction. This happens along directions oriented at $\pi/4$ with respect to one another for $h_\oplus$ and $h_\otimes$ polarizations. Image credit *Eur Phys J Plus* 132 (2017) no 1,10.

by considering a dissipative spring with masses m_1 and m_2 separated at an equilibrium distance, L, with spring constant, k [1] Like electromagnetic waves, GWs are believed to have two polarizations, $\oplus$ and $\otimes$. These indicate the distortion that a ring of particles assume, as shown in Fig (6.3) Suppose that a $\oplus$-polarized GW collides with the bar, and induces damped oscillations

$$\ddot{x} + 2\beta\dot{x} + \omega_0^2 x = -\frac{1}{2}h_\oplus L\omega^2 \cos\omega t,$$

with an unperturbed frequency, $\omega = \sqrt{(k/\mu)}$, where μ is the reduced mass. Applying energy conservation, the work against the action of dissipative forces, $\beta\dot{x}^2$ comes from the decrease of the total energy, $\frac{1}{2}d/dt(\mu\dot{x}^2 + \omega_0^2 x^2)$. That is, the loss of the energy

---

[1]W Anderson & J Creighton, *Gravitational Wave Physics and Astronomy: An Introduction to Theory* (Wiley, New York, 2012).

contained in the GW shows up as 'heat.' Moreover, through this energy loss, transverse GWs are converted in longitudinal acoustic waves.

Doesn't the lack of discrimination of GR permit us to choose whatever frame appears to us more convenient? Undoubtedly, attaching coordinates directly to test particles in free-fall seems the most convenient. But choosing the transverse traceless (TT) frame means that the test particles must necessarily follow geodesics which are the counterpart of Euclidean inertial frames, and hence there can be no change of energy when a GW collides with a bar, or whatever else gets in its way.

How much of the results that are obtained depend on the choice of gauge that is employed. One may, if one desires, choose to decompose the metric into radiative TT, gravitational, and unphysical gauge components. [2] Only the TT-component of the metric will satisfy a wave equation and supposedly propagate at the speed of light, while the other components satisfy an equation of the Poisson type. However, just because the free-fall component satisfies a wave equation doesn't imply that it will radiate. [3]

Things like energy and energy fluxes only become clear in the Newtonian approximation of GR for 'weak' fields. Then what allows us to go beyond this approximation where energy-momentum is not localized, nor even conserved?

Since GWs knows no shielding and should affect everything in their path, should it also affect the spring constant? As the GW passes, the spring should also be effected and loosen up. Cooperstock and Tieu [4] argue precisely this: "what has been overlooked is that the bar itself has been presumed unaffected by the gravity waves."

However, for the sake of argument, let us assume that the loss in energy of a GW shows up in heat as a consequence of energy conservation. As Duerr [5] remarks, energy and momentum conservation follow from the homogeneity of time and space, respectively, and there is nothing to guarantee that such symmetries hold in general space times that

---

[2] E E Flanagan & S A Hughes, "The basics of gravitational wave theory," arXiv/gr-qc/0501041 (2005).

[3] Duerr, "Do gravitational waves carry energy?-Critique of a procrustean practice"

[4] F Cooperstock & S Tieu, *Einstein's Relativity: The Ultimate Key to the Cosmos* (Springer, New York, 2012).

[5] P M Duerr, *op. cit.*

accommodate GWs. Schroedinger [6] summed things up by saying:

> the total amount of energy or mass contained in a closed expanding uni-
> verse *decreases*. In simple models the loss can be computed and equals the
> amount of work the *pressure* would have to do to increase the volume, if a
> piston had to be pushed back as in the case of an adiabatically expanding
> volume of gas. Yet there is nothing like a piston nor any boundary at all
> through which the energy could escape.

How far down the macroscopic-microscopic scale must you go to analyze the stick
bead argument? And if GR fails to conserve energy why should Feynman's sticky
bead argument be made to do so?

So one must admit that in order for the sticky beads to heat up, the particles in the
bar must no longer follow geodesic paths. Then energy is lost and the beads heat up.
Either GR is unable to account for such phenomena, or the books don't balance and
one has to admit that there is no conservation of energy-momentum in GR.

One then would lay the blame on the non-Euclidean nature of GR, and all non-
Euclidean theories of gravitation will also manifestly not conserve energy-momentum.
In GR, the covariant divergence of the sum of the energy-stress tensor and the pseudo-
tensor vanishes. However, according to Weyl [7]

> Indeed all the [pseudo-tensor components] can through a suitable
> choice of a coordinate system be made to vanish ... on the other hand,
> one obtains [pseudo-tensor components] that are different from zero in a
> 'Euclidean', completely gravity-free world by using a curvilinear coordi-
> nate system.

So the blame cannot be laid on the non-Euclidean nature of GR. If we impose the
restriction that the coordinates should be Cartesian we are breaking with a basic tenet
of GR in that there should be no privileged reference frames. How much of the results

---

[6]E. Schroedinger, *Space-Time Structure* (Cambridge U P, Cambridge, 1950) p 105.
[7]H Weyl, *Space-Time-Matter*, 4th ed (Dover, New York, 1952).

of GR are dependent upon the background metric, and the *local* nature of pseudo-tensors that don't exist in general spacetimes?

6.2 Why are GWs transverse?

Ampère's investigation were based on the assumption that the force acting between two current elements is directed along the line connecting them.

Maxwell [8] attempted to generalize the force on ds arising from the action of ds' in a way "consistent with experimental facts." He assumed that the force acting on ds is the sum of three components: one in the direction of r:

$$d^2F = \left\{ \frac{1}{r^2}\left(\frac{\partial r}{\partial s}\frac{\partial r}{\partial s'} - 2r\frac{\partial^2 r}{\partial s \partial s'}\right) + r\frac{\partial^2 Q}{\partial s \partial s'} \right\} ids \cdot i'ds',$$

and components in the direction of ds and ds',

$$d^2S = -\frac{\partial Q}{\partial s'}ids \cdot i'ds',$$

and

$$d^2S' = \frac{\partial Q}{\partial s}ids \cdot i'ds',$$

respectively.

We only know that Q is some function of r. Therefore, we have to resort to experiment to determine its nature. Those experiments consist in current forming closed loops. If Ampère is correct that the action between elements ds and ds' is along a line joining them the S and S' terms must be zero, and Q must reduce to a constant, or zero. What is left is the negative of Ampère's force.

Now Grassmann assumes the opposite: the two elements have no mutual action along the line connecting them, but, rather:

$$Q = -\frac{1}{2r}, \quad F = -\frac{3}{2r}\frac{\partial^2 r}{\partial s \partial s'}, \quad S = -\frac{1}{2r^2}\frac{\partial r}{\partial s'}, \quad S' = \frac{1}{2r^2}\frac{\partial r}{\partial s}.$$

---

[8] *op. cit.* §535 p 160.

Choosing between these alternatives, Maxwell concludes, that is only Ampère's law which satisfies Newton's III of equal action and reaction. And it is this force which is most frugal in its explanation since it does not require the introduction of transverse fields to the direction of the motion.

We recall that Weber's force, which is the time analog of Ampère's law was used in the study of the advance of the perihelion of Mercury and the deflection of light. Is there a distinction between the gravitational force between two bodies and GWs? One could try to formulate a theory in which the gravitational force propagates radially, but the fields are transverse to the direction of propagation. But, then, one would have to specify what those fields are. One could also argue that Newton's law like Coulomb's law acts instantaneously and separate it from Maxwell's laws and the connection between electric and magnetic fields and electromagnetic radiation. But, we have no gravitational analogue of Maxwell'equation, nor can we since one is tensorial in character while the other vectorial. That is the reason why GR always has to go to the weak field, or asymptotic, limit to the reintroduce Newtonian concepts of potential and force.

It is precisely for this reason that Ampère and Weber lost out to Grassmann when relativity came onto the scene. Yet we have no idea what the gravitational analogues are of the electromagnetic fields, and can't even localize energy in GR much less explain how gravitational radiation occurs.

How then can one so assuredly claim that GWs are transverse, and the polarization they are credited to have? They are forced to borrow properties from their EM cousins. Returning to Feynman's sticky bead argument, couldn't he also explain the existence of an electromagnetic radiation pressure in exactly the same way?

6.3 Free-fall in a gravitational field?

Conventional wisdom asserts that the "universality of free-fall motivates the geometric interpretation of gravity, with small test masses moving along geodesic coordinate of a curved spacetime geometry.[9] This lies at the very heart of what is wrong with GR. Ohanian goes on to state that in GR, "this geodesic motion can be shown to be a consequence of the 'conservation' law $T^\nu_{\mu;\nu} = 0$[10] for non-gravitational matter." He then argues that this is really not a conservation laws, but rather a 'conservation' law since

> it reveals to what extent energy-momentum of the nongravitational matter is *not* conserved. Written out in full, it becomes
>
> $$\frac{\partial}{\partial x^\mu} T^\mu_\nu = \Gamma^\alpha_{\nu\mu} T^\mu_\alpha - \Gamma^\mu_{\alpha\mu} T^\alpha_\nu \qquad (6.1)$$
>
> which determines the rate at which non-gravitational matter receives energy and momentum from the non-gravitational field...Only in a local geodesic coordinate frame (that is, in freely falling coordinates with $\Gamma^\alpha_{\mu\nu} = 0$) does Eq[(6.1)] reduce to the standard form of a conservation law, $\partial T^\mu_\nu / dx^\mu = 0$, which shows that the gravitational field delivers no energy or momentum of a small test mass...are constant, so the body remains at rest in the freely falling coordinates and therefore moves along a geodesic.

What Ohanian is saying is that gravity resides in $\Gamma^\alpha_{\nu\mu}$, and the fact that the particle moves along a geodesic is not sufficient, it must move in Euclidean space for there to be conservation. Gravity if embodied in a force creates acceleration so that a particle acted upon by gravity will not move along a geodesic. And even if it does move along a geodesic the fact that $\Gamma^\alpha_{\mu\nu} \neq 0$ does not mean that gravity is present.

As we have already noted, the Christoffel symbols, (5.18) contain first-order derivatives of the metric tensor. These first-order derivatives can be made to vanish whereas second-order ones cannot, and the Riemann tensor deals exclusively with the latter. So, in contrast to Ohanian,

$$\Gamma^\alpha_{\mu\nu} \frac{dx^\mu}{d\tau} \frac{dx^\nu}{d\tau} \neq 0$$

is not the 'doorway' for letting gravity into an equation that says the proper acceleration is zero.

---

[9]H Ohanian, "The energy-momentum tensor in GR, and alternative theories of gravitation and the gravitational vs inertial mass" arXiv:1010.5557[gr-qc].

[10]The semicolon indicates the covariant derivative.

In other words, the Christoffel symbols, (5.18) are not tensors, just like the pseudo-tensor which is made up of products of them, so they can be made to vanish in a selected coordinate system, like Fermi coordinates for a time-like geodesic.

In short, Ohanian would have us believe that all non-Euclidean geometries are gravity, based on the misconception that gravity resides in the distortion of Euclidean geometry.

It is also astonishing that the GR proponents can't get the story straight among themselves. In a recent paper that purports a 'simplified' derivation of the gravitational stress tensor, [11] the author linearizes the field equations and then writes the first-order contribution as:

$$G^{(1)}_{\mu\nu} = -8\pi G \left(T_{\mu\nu} + t_{\mu\nu} \right), \tag{6.2}$$

where $G^{(1)}_{\mu\nu}$ consists of terms that are linear in the perturbed metric, $h_{\mu\nu}$, and

$$t_{\mu\nu} = \frac{1}{8\pi G} \left(G^{(2)}_{\mu\nu} + \cdots \right),$$

where the second-order Einstein tensor represents quadratic terms in $h_{\mu\nu}$. $t_{\mu\nu}$ is referred to as a pseudo-tensor, but the "sum $T_{\mu\nu} + t_{\mu\nu}$ is often referred to as the energy-momentum pseudo-tensor." However, in a post on the internet, "Strange misconceptions of general relativity," 't Hooft, emphasizes that "any modification of Einstein's equations into something like

$$R_{\mu\nu} - (1/2)R g_{\mu\nu} = \kappa \left(T_{\mu\nu}(\text{matter}) + t_{\mu\nu}(\text{gravity}) \right)$$

where $t_{\mu\nu}$ would be something like a 'gravitational contribution' to the stress-energy-momentum tensor, is blatantly wrong." He then goes on to state

> Writing such a proposal betrays a complete misunderstanding of what General Relativity is about. The energy and momentum of the gravitational field is completely taken into account by the non-linear parts of the original equation. . .Note that a freely falling observer experiences no gravitational field and no energy-momentum transfer; hence there cannot be a covariant tensor such as $t_{\mu\nu}(\text{grav})$

---

[11] S Blabus, "Simplified derivation of the gravitational wave stress tensor from the linearized Einstein field equations," *PNAS* **113** (2016) 11662-11666.

First, the pseudo-tensor, $t_{\mu\nu}$(gravity) is not a tensor because it contains only first derivatives in the $h_{\mu\nu}$. Second, a freely falling observer follows geodesic paths with $\Gamma^\alpha_{\mu\nu} \neq 0$, so Ohanian wouldn't agree with him. Third, Balbus opts for a 'bootstrapping' iteration scheme, where 'gravitational energy' should be the source of the 'gravitational field.' In this scheme, the first-order Einstein tensor is given by the sum of the energy-stress tensor and pseudo-tensor where the pseudo-tensor is given by the second-order Einstein tensor, and so on. If such a scheme would work, it would represent an effective force field theory laid on a Minkowski background.

Finally, Balbus cannot write down any balance equations whatsoever without introducing an 'effective' potential that satisfies the "wave equation of scalar gravity." So that without recourse to good-old Newtonian concepts, there are no energy densities and energy fluxes.

If GWs are caused by accelerating masses, it is not for the field equations to pronounce upon. To bring this point home, we can compare term by term in the geodesic equation:

$$\frac{d^2 r}{d\tau^2} + \Gamma^r_{tt}\left(\frac{dt}{d\tau}\right)^2 + \Gamma^r_{rr}\left(\frac{dr}{d\tau}\right)^2 + \Gamma^r_{\varphi\varphi}\left(\frac{d\varphi}{d\tau}\right)^2 = 0,$$

in the plane $\vartheta = \pi/2$. We now want to compare this equation with Weber's gravitational equation with $k = -1$, for the Schwarzschild field in the *weak* field limit.

In this limit, the geodesic equation are:

$$\frac{d^2 r}{d\tau^2} - r\left(\frac{d\varphi}{d\tau}\right)^2 = -g\left(1 - \frac{1}{c^2}\left(\frac{dr}{d\tau}\right)^2\right), \tag{6.3}$$

and

$$r\frac{d^2\varphi}{d\tau^2} + 2\frac{dr}{d\tau}\frac{d\varphi}{d\tau} = 0.$$

Comparing the radial geodesic equation, (6.3), the Weber equation of gravitation with $k = 1$,

$$\frac{d^2 r}{d\tau^2} - r\left(\frac{d\varphi}{d\tau}\right)^2 = -\frac{GM}{r^2}\left(1 - \frac{1}{c^2}\left(\frac{dr}{d\tau}\right)^2 + \frac{1}{c^2}r\frac{d^2 r}{d\tau^2}\right),$$

we clearly notice the absence of the acceleration term on the right-hand side of the former. This affirms that geodesics always have zero tangential accelerations.

Particle accelerations are as alien to the metric $d\tau^2 = g_{\mu\nu}dx^\mu dx^\nu$ as accelerations are to Doppler shifts. The acceleration term in the Weber force originates in the interaction of the two current elements.

The acceleration corresponds to the second derivative in two current elements. When taken over a closed circuit, this vanishes and leads to the equivalence of the Neumann and Weber electromagnetic energies which would otherwise not be equivalent. That is to say that the electrodynamic energy, according to Neumann, is:

$$\frac{ii'(d\vec{s}\cdot d\vec{s}')}{r},$$

while, according to Weber, it is:

$$\frac{ii'}{r}(\hat{r}\cdot d\vec{s})(\hat{r}\cdot d\vec{s}').$$

These expressions differ by:

$$\frac{ii'dsds'}{r}\left\{\cos(r\cdot ds)\cos(r\cdot ds') - \cos(ds\cdot ds')\right\},$$

or, equivalently by the acceleration,

$$ii'dsds'\frac{\partial^2 r}{\partial s\partial s'},$$

which when integrated round either circuit vanishes.

A general formula that includes both Neumann and Weber is: [12]

$$k\frac{ii'(d\vec{s}\cdot d\vec{s}')}{r} + \frac{\partial^2 r}{\partial s\partial s'}dsds',$$

where k is an arbitrary constant. We have chosen $k = -1$ in Weber's gravitational force law.

This underscores a very important point when discussing the possibility that the Schwarzschild field can admit 'positive' acceleration, or, equivalently, gravitational repulsion. The geodesic equation:

$$\ddot{x}^j + \Gamma^i_{jk}\dot{x}_j\dot{x}_k = 0, \tag{6.4}$$

---

[12]H Lamb, *Proc Lond Math Soc* **xiv** (1883) 301.

insures us that the acceleration in the tangent plane to the given surface is zero.

Consider the parametric form of a curve, $\alpha = h(x(\tau))$. If τ is time then

$$\dot{\alpha}_i = h_{i,j}\dot{x}_j \tag{6.5}$$

is the velocity where the comma stands for differentiation, and there is a sum over repeated indices. This must be constant in magnitude but it can change its direction. Multiplying (6.5) by $\dot{\alpha}_k$ results in:

$$\dot{\alpha}_k\dot{\alpha}_k = h_{k,i}\dot{x}_i h_{k,j}\dot{x}_j = g_{ij}\dot{x}_i\dot{x}_j$$

which must also be constant, where the metric has been defined as [13]

$$g_{ij} = h_{k,i}h_{k,j}. \tag{6.6}$$

The condition that the velocity,

$$V = \sqrt{\left(g^{ij}\dot{x}_i\dot{x}_j\right)}, \tag{6.7}$$

is constant will yield the geodesic equations as being the shortest paths to given end-points:

$$\int_{\tau_1}^{\tau_2} V(\dot{x}, x)d\tau. \tag{6.8}$$

Not only should (6.8) be an extremum, the condition of constant velocity, $\dot{V} = 0$ is added to get a unique solution. It can be shown [14] that the equations, resulting from requiring that (6.8) be an extremum,

$$2\frac{d}{d\tau}\left(g^{ij}x_j\right) - \frac{\partial}{\partial x_j}g^{ij}\frac{dx_i}{d\tau}\frac{dx_j}{d\tau} = 0, \tag{6.9}$$

satisfy the condition $\dot{V} = 0$. Equations (6.9) are identical to the geodesic equations, (6.4).

A second differentiation in time,

$$\ddot{\alpha}_i = h_{i,jk}\dot{x}_j\dot{x}_k + h_{i,j}\ddot{x}_j \tag{6.10}$$

---

[13]P. Pokorny, "Geodesics Revisited," *CMSIM* (2012) 281-298.
[14]Pokorny, *op. cit.*

gives the accelerations.

If we now impose the condition that component of the acceleration in the plane tangent to the curve be zero we get the geodesic equation:

$$\ddot{\alpha}_k r_{k,j} = 0, \qquad \forall i. \tag{6.11}$$

The geodesic is determined by $g_{\mu\nu}$ and its first derivatives only. It can tell us nothing about the curvature of the surface because, for that, you need also the second derivatives of the metric. The fact that the pseudo-tensor is comprised only of products of the Γ^i_{jk} would make gravity the equivalent to a geodesic equation. If it so happens that the metric is constant so that all the Christoffel symbols vanish, $\ddot{x}_i = 0$, the geodesic becomes a straight line.

Therefore, the second term in (5.17) is **not** the negative of the acceleration. There is only one component of the acceleration that persists, and that component is normal to the surface; the component of the acceleration lying in the tangent plane is zero. The obvious example that comes to mind is uniform rotational motion, which in hyperbolic space, is described by the Beltrami metric. [15]

This says a lot about the literature. Most recently, with the inflationary scenario, [16] a positive acceleration was taken to mean repulsion with the consequence that the universe is expanding. A detailed list of authors claiming that the Schwarzschild field exhibits repulsion can be found in the McGruder [17] paper that we discussed earlier. There is only one acceleration possible and that is normal to the surface for geodesic paths.

We can even say more about what is permissible in the geodesic equation if we consider the optical geodesic equation:

$$\ddot{r} - r\dot{\varphi}^2 = \nabla n^2, \tag{6.12}$$

For isotropic, inhomogeneous systems, the index of refraction is a decreasing function of r. By increasing the power of r, n^2 varies from $1/r$, an ellipse, $1/r^2$, Cotes' spiral,

---

[15]Lavenda, *op. cit.*, p 445.

[16]Guth, *op. cit.*

[17]*op cit.*

to $1/r^3$, which is Einstein's modification for the deflection of light. And together with $1/r^2$ it gives the advance of the perihelion.

Now, the index of refraction cannot be a function of the speed for geodesic motion. Take for instance, an index of refraction of the form:

$$n^2 = \frac{c^2 - \dot{r}^2}{r}.$$

Introducing this into (6.12), gives:

$$\ddot{r} - r\dot{\varphi}^2 = -\left(\frac{c^2}{r^2} - \frac{\dot{r}^2}{r^2} + 2\frac{\ddot{r}}{r} \right). \tag{6.13}$$

We immediately recognize the right-hand side as the negative of Weber's force . As we mentioned earlier, (6.13) is Gerber's equation for the advance of the perihelion to within a scale factor of $3\mu/c^2$ for the Weber force. We can now appreciate it that it does *not* correspond to geodesic motion. In contrast, Einstein's modification of the equation of the orbit, (2.78), is a geodesic equation.

6.4 A history of GWs

The recent 'discovery' of GWs by the LIGO/Virgo team [18] not only confirms the existence of GWs, but, moreover, adds credence to black holes (BH), and shows that binary BHs may collide, which during their merging process emit GWs.

All this is in face of the presumed isolation of a BH by its event horizon. If nothing can escape the event horizon because nothing can propagate faster than the speed of light how can the gravitational field get out of a BH so as to create a binary pair?

Van Flandern [19] asked the same question, and never got an answer. He pictorially

---

[18]B P Abbott, *et. al.* "Observation of Gravitational Waves from a Binary Black Hole Merger," textitPhys Rev Lett **116** (2016) 061102.
[19]*op. cit.*

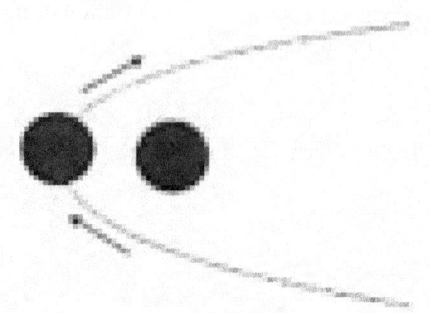

Figure 6.4: Van Flandern asked the question on how can black holes communicate when their masses are hidden behind event horizons that act as 'fire-walls.' Taken from T Van Flandern, "The speed of gravity: What the experiments say."

envisaged the BHs in Fig (6.4). In fact, in the Russian literature predating the coining of the term BH, such stars were known as 'frozen' stars.

The existence of GWs was predicted by Einstein in 1916 using linearized field equations by slightly perturbing the metric of Minkowski, $\eta_{\mu\nu}$, by adding on a 'small' perturbation:

$$g_{\mu\nu} = \eta_{\mu\nu} + \epsilon h_{\mu\nu}, \tag{6.14}$$

where ϵ is a small, positive number much less than 1. But, we are not told small in compared to what.

Introducing (6.14) into his field equations and discarding terms in ϵ^2 and higher, Einstein obtained: [20]

$$\left(\frac{\partial^2}{\partial t^2} - \nabla^2 \right) \tilde{h}_{\mu\nu} = \kappa T_{\mu\nu}, \tag{6.15}$$

where $\kappa \equiv 8\pi G$, and $\tilde{h}_{\mu\nu} = h_{\mu\nu} - (1/2)\eta_{\mu\nu}h$.

The energy-stress tensor, $T_{\mu\nu}$, is purposely given the impression of being the source of GWs, even though it does not contain gravitational energy, for it is claimed that "out-

---

[20] C D Hill & P Nurowski (HN)

side the sources where $T_{\mu\nu} = 0$, [(6.15)] constitute a system of decoupled relativistic wave equations,"

$$\left(\frac{\partial^2}{\partial t^2} - \nabla^2\right)\tilde{h} = 0. \tag{6.16}$$

Eddington [21] makes a more admiral attempt of writing (6.15) as:

$$\left(\frac{\partial^2}{\partial t^2} - \nabla^2\right)\tilde{h}_{\mu\nu} = \kappa[G_{\mu\nu}], \tag{6.17}$$

being noncommittal by replacing the 'source' by the Einstein tensor $G_{\mu\nu}$, where the brackets denote that it has "prescribed values at every point."

According to HN, and just about everyone else, the free wave equation, (6.16), "enabled Einstein to conclude that *linearized* GR theory admits solutions in which the perturbations of Minkowski spacetime, $h_{\mu\nu}$, are plane waves traveling with the speed of light."

Eddington had a different take on the matter entirely. His solution to (6.17) is:

$$\tilde{h}_{\mu\nu} = \frac{1}{2\pi r}\{G\}_{t-r} = \frac{A_{\mu\nu}}{2\pi}\frac{e^{i\omega(t-r)}}{r}, \tag{6.18}$$

where the curly brackets denote a triple integral over space, and the subscript means that it is to be evaluated at the retarded time, $t - r$.

Eddington refers to his solution as "the usual formula for spherical waves in the theory of sound and electromagnetism." But this is not all; Eddington continues:

> I am not sure that any writer has explicitly stated that gravitation is propagated isotropically by spherical waves of this kind; but I think that many like myself must have regarded it as an inference from Einstein's equation which was two [sic] obvious to need mentioning. It was a great surprise to me to find that the solution (6.18) *does not satisfy the law of gravitation.*

---

[21]A S Eddington, "The propagation of gravitational waves," *Proc Roy Soc A* **102** (1922) 268-282.

He is forced to conclude that the only way to satisfy (6.18) is in "the static case of an undisturbed particle. . .since isotropic spherical waves of gravitation cannot occur." What then is the purpose of introducing the perturbation (6.14) in the first place?

HN then go on to say that due to superposition, "one can get waves having any desirable wave front. . . [Einstein] also showed that within the linearized theory these waves carry energy, and he found a formula for the energy loss in terms of the third time derivative of the quadrupole moment of the sources."

Again, Eddington has a different story to tell, and he tells it this way:

> The chief point of interest in the problem of the spinning rod is the question of whether its energy is gradually carried away by GWs which are created, so that the rotation would slow down and ultimately stop of its own accord.

However, he doesn't use his solution, but, rather, introduces "the pseudo-energy-tensor of the gravitation field, correct to second order," referencing Einstein's second paper on the subject. It is a 'pseudo'-tensor because it consists of bilinear products of the first derivatives of the perturbed metric, just like the Christoffel symbols.

Previous to this he calculated the components of the energy stress tensor. For a rod of length $2a$ with mass σ per unit length, having angular speed ωa, all non-vanishing contributions to the energy-stress tensor have magnitudes $(2/3)\sigma\omega^2 a^3$. Why then did he find it necessary to introduce the fictitious pseudo-tensor? Eddington, himself, refers to the use of the pseudo-tensor as mere 'fiction.'

Yet, he does so in the calculation of the Poynting vector which he associates with $4\pi r^2$ times the mixed time and space components of the pseudo-tensor, t_t^r along the radius. When averaged over the sphere of radius r, Eddington finds that energy is radiated at a steady rate of $(1/10)[(2/3)\sigma\omega^2 a]^2$, or in terms of the moment of inertia I of the rod, $(32/5)I^2\omega^6$. This is twice as much as Einstein found, and "[t]he discrepancy is due to a numerical slip in one or the other investigation, and is not of much importance."

	PSR1913+16	PSR1534+12
a/c (sec)	2.342	3.729
P (sec)	27,907	36,352
$\dot{P}$ -observed	-2.42x10^{-12}	±0.6x10^{-12}
$\dot{P}$ -predicted	+921x10^{-12}	+1682x10^{-12}

Figure 6.5: Comparison of the rates of change in the periods of binary pulsars PSR1913+16 and PSR1534+12 with those calculated from (5.6). The data was taken from J H Taylor, A Wolszczan, A Damour & J M Weisberg, "Experimental constraints on strong-field relativistic gravity," *Nature* **355** (1992) 132-136. The Table was prepared by Van Flandern *op. cit.*

Yet, what is of importance is how can "resolving the [mixed component of the pseudo-tensor] along the radius" ever give "the radial component of the flow of energy per unit time (Poynting vector)"? Did Eddington have a lapse of memory that by a mere change of coordinates the pseudo-tensor can be made to vanish? Energy densities and fluxes cannot be made to vanish by a mere wave of a magic wand.

6.5 Dissipative binaries

Feynman's analogy between GW radiation and sticky beads presumably acquired respectability in the wake of the Hulse-Taylor [22] (HT) discovery of the binary pulsar PSR 1913+16 in 1974. HT found that the period of the binary was declining which, supposedly, implied that the pair was rotating faster and faster in an ever tightening orbit. The corresponding decline of the orbiting period was measured to be 75 millionths of a second per year.

Van Flandern [23] prepared a table shown in Fig (6.5). He concludes that there is no possible match between observed rates and those predicted from (5.6) that claims gravity propagates at the speed of light. The observed values are orders of magnitude larger than the observed values and for PSR1913+16 they are even of opposite signs as well. From the data on PSR1534+12, Van Flandern sets a lower limit on the speed of gravity to be 3 thousand times larger than the speed of light.

---

[22] R A Hulse & J H Taylor, "Discovery of a pulsar in a binary system," *Ap J* **195** (1975) L51-L53.
[23] *op. cit*

6.5.1 Qualms concerning GWs

Is GR the right venue for discussing dissipative binaries? In GR: [24]

▨ there is no solution of Einstein's equations for a field of two particles or singularities;

▨ there are no bound solutions to the two body problem; [25]

▨ there are no periodic solutions to the Einstein field equations. Consequently, since orbits are unstable by their very nature, it is useless to attempt to explain their instability by an energy loss mechanism; [26]

▨ there is no principle of superposition, because the sum of solutions will not satisfy $G_{\mu\nu} = 0$;

▨ there is no possibility of uniform rotation without radiation, for according to Eddington it is begging the question to first constrain the bodies to rotate uniformly, and then "calculate the resulting gravitational waves, and verify that the radiation of gravitational energy across an infinite sphere is zero";

▨ the metric $g_{\mu\nu}$ is extremely sensitive to coordinate changes. In Eddington's view coordinate changes can be propagated at "*the speed of thought*," and "these may be mixed up at will with the more dilatory propagation" that he considered GWs to propagate at. For if one set of coordinates are chosen, "the speed is that of light; if the coordinates are slightly different the speed is altogether different from that of light."

▨ there is absolutely no way that $g_{\mu\nu}$ can be decomposed into conservative and dissipative components—into a 'flat' background with 'ripples' of spacetime superimposed on it.

What is not true, or hasn't been shown, contrary to claims that have been made are that: [27]

▨ GWs are vibrations in spacetime propagating at the speed of light;

---

[24] A S Eddington, *The Mathematical Theory of Relativity* (Cambridge UP, Cambridge, 1923) pp 95, 130.

[25] A Papapetrou,"Ueber periodische nicht-singulare Losungen in der Allgemeinen Relativitatstheorie," *Ann Phys* **20** (1957 399.

[26] Duerr, *op. cit.*

[27] L Bieri, D Garfinkle and N Yunes, "Gravitational waves and their mathematics," in *The Mathematics of Gravitational Waves*, Part 2, p 693.

▓ the Schwarzschild spacetime describes a BH;

▓ GWs travel along null geodesics, just like light waves. "Geodesics that enter the event horizon of a BH never leave. Objects in free fall travel along time-like geodesics... also can never leave once they have entered a BH;

▓ all particle trajectories follow geodesics – "curves with 0 acceleration;"

▓ "when two BHs fall into each other, they merge and form a single larger BH."

If light can't escape from a BH neither can GWs. Where is the gravitational attraction needed to keep the two binaries in orbit, and why should they attract each other once their distance exceeds the sum of their event horizons?

The Schwarzschild solution can neither be extended in the region smaller than the Schwarzschild radius, nor can the singularity in the metric be displaced to the origin by discriminating between 'inessential' and 'essential' singularities. Coordinate transforms won't work on non-Euclidean metrics without changing them entirely so why should they work on the Schwarzschild metric? And why should we ignore Schwarzschild's inner solution in favor of extending his outer solution beyond its boundaries?

If all particle trajectories are constrained to geodesics, which are the next best thing to inertial motion in Euclidean space, how do you explain all the cataclysmic events predicted from Numerical Relativity (NR)? If there are no forces in GR how can there be potentials from which they are derived? And without potentials how can one seriously speak of energy, its localization, and its propagation?

And then there is the nagging question on how energy loss, through radiation or by any other means, can be analyzed within Einstein's field equations, $G = \kappa T$, when the left-hand side also demands that

$$\nabla \cdot T = 0? \tag{6.19}$$

One also needs to specify boundary conditions, i.e., that Einstein's equations "satisfy a *condition for absence of incoming radiation*." [28]

---

[28] J Ehlers, A Rosenblum, J N Goldberg & P Havas, (ERGH) "Comments on gravitational radiation damping and energy loss in binary systems," *Ap J* **208** (1976) L77-L81.

How then can one hope to explain radiation losses in an isolated system when one is confronted with something like (6.19)? What one normally does it to impose a coordinate condition on $g_{\mu\nu}$ that results in a "relaxed field equation" having the same form as Einstein's original equation but without the conservation condition (6.19). [29]

Both Einstein's original derivation of his GWs based on a linear approximation, as well as all those who faithfully reproduce it [30] fail to take into account the restrictions on the motion imposed by the field equations. [31]

Moreover, a radiation-reaction force, and the work in overcoming it, are also completely foreign to, and outlawed by Einstein's equations. They, in fact, allow only one type of motion for given initial conditions. ERGH sum up the situation nicely when they say: "Violating the equations of motion is analogous to considering solutions of Maxwell's equation that violate the equation of continuity."

6.5.2 "Derivation" of radiated power

Specifically, what is done is to return to Eddington's solution (6.18), and to replace the Einstein tensor by τ_i^k, something that derives from the energy-stress tensor, but comprises of "auxiliary quantities which are obtained upon going over from the exact equations of gravitation to the case of a weak field." [32] Reinstating c, Eddington's equation becomes:

$$\tilde{h}_i^k = -\frac{\kappa}{R_0}\int [\tau_i^k]_{t-R_0/c}dV, \tag{6.20}$$

where R_0 is the distance from the body of mass m to the origin. [33]

Now, instead of T satisfying (6.19), $\tau_{\mu\nu}$, is required to satisfy the *conservation* equations:

$$\frac{\partial \tau_{\mu\nu}}{\partial x^\nu} - \frac{\partial \tau_{\mu 0}}{\partial x^0} = 0, \qquad \frac{\partial \tau_{\mu 0}}{\partial x^\mu} - \frac{\partial \tau_{00}}{\partial x^0} = 0, \tag{6.21}$$

which represent the conservation of momentum and continuity equations, respectively.

---

[29]ERGH, *op. cit.*

[30]L D Landau & E M Lifshitz, *Classical Theory of Fields* (Addison-Wesley, Reading MA, 1951), ch 9; S Weinberg, *Gravitation & Cosmology* (Wiley, New York, 1972); H C Ohanian, *Gravitation & Spacetime* (Norton, New York, 1976).

[31]A Loinger & T Marisco, "All relativistic motions can be geodesically described," arXiv:1006.3844.

[32]Landau & Lifshitz, *op. cit.*.

[33]Since GR does not deal with forces and potentials how can you retard something that isn't there?

Multiplying the first by x^λ and integrating over a volume V, and multiplying the second equation by $x^\lambda x^\xi$, and again integrating the result over the volume, performing an integration by parts, where the boundary terms 'vanish at infinity' (requiring an integration over an infinite volume) and comparing results give:

$$\int \tau_{\mu\nu} dV = \frac{1}{2} \frac{\partial^2}{\partial x_0^2} \int \tau_{00} x^\mu x^\nu dV,$$

where any and all vestiges of retardation have disappeared.

Considering τ_{00} to be the rest energy of a particle of mass M [34],

$$\tau_{00} = Mc^2, \tag{6.22}$$

and with $t = x^0/c$, there results:

$$\tilde{h}_{\mu\nu} = -\frac{\kappa}{4\pi R_0} \frac{\partial^2}{\partial t^2} \int M x^\mu x^\nu dV, \tag{6.23}$$

not forgetting the fact that we are dealing with an infinite volume.

In the next breadth, Landau & Lifshitz want to "calculate the flux of energy radiated by the system using the expression:

$$t^{ik} = \frac{1}{4\kappa} \tilde{h}_q^{n,i} \tilde{h}_n^{q,k}, \tag{6.24}$$

for the pseudo-tensor, leaving off the adjective, energy-momentum, because it isn't. Here, there is much to criticize:

■ Some sort of conservation laws, (6.21), are used to calculate, a vestige of, the energy-stress tensor, $\tau_{\mu\nu}$, only to have us use it to calculate the energy radiate by GWs in an *infinite* volume. Where is the observer supposed to be placed? on the surface to measure the Poynting energy flux?

■ After having defined what remains of an energy-stress tensor, why resort to the 'fiction', as Eddington put it, of a pseudo-tensor, which can neither represent energy nor the flux of energy?

---

[34]It will be appreciated in the sequel that this 'rest' energy is confused with internal energy.

◼ Since the rest energy has alone been used in the energy-stress tensor, the mass must be a rest, and, supposedly, masses at rest neither generate GWs, nor radiate energy!

The final step is to associate a GW with an electromagnetic wave with two independent polarizations. If the 'energy flux' is directed along the x^1 direction, the two components are:

$$\tilde{h}_{23} = -\frac{\kappa}{12R_0}\ddot{Q}_{23}, \qquad \tilde{h}_{22} - \tilde{h}_{33} = -\frac{1}{12R_0}\left(\ddot{Q}_{22} - \ddot{Q}_{33}\right), \qquad (6.25)$$

where $Q_{\mu\nu}$ is the mass quadrupole tensor:

$$Q_{\mu\nu} = \int M\left(3x_\mu x_\nu - \delta_{\mu\nu}r^2\right)dV.$$

Thus, the pseudo-tensor (6.24) is:

$$t^{10} = \frac{\kappa}{122\pi^2 c^2 R_0^2}\left[\left(\frac{(d^3/dt^3)Q_{22} - (d^3/dt^3)Q_{33}}{2}\right)^2 + (d^3/dt^3)Q_{23}^2\right],$$

which is **not** "the flux of energy into an element of solid angle" when multiplied by the 2-sphere, $r^2 do^2$.

Integrating over the sphere of infinite volume, results in a radiated power loss of:

$$\frac{d\mathcal{E}}{dt} = -\frac{G}{45c^5}\left(\frac{d^3 Q_{\mu\nu}}{dt^3}\right)^2 = -\frac{2}{5}\frac{GM^2}{a^2}(a\omega)\left(\frac{a\omega}{c}\right)^5, \qquad (6.26)$$

where, in the last equality, r has been replaced by a, the distance between two equal mass stars $M/2 = m_1 = m_2$, and ω is the angular speed of either star.

The detailed calculation was performed by Peters & Mathews, [35] and carried over *verbatim* by HT. [36] In order to determine the luminosity from (6.26), we have to introduce Kepler's III, then by multiplying numerator and denominator by c^2 we come out with:

$$\mathcal{L} = \frac{2}{5}Mc^2\left(\frac{a\omega}{c}\right)^7\omega, \qquad (6.27)$$

[35] P C Peters & J Mathews, "Gravitational radiation from point masses in a Keplerian orbit," *Phys Rev* **131** (1963) 435-440.

[36] J M Weisberg & J H Taylor, "The relativistic binary pulsar B1913+16: Thirty years of observation analysis," in *Binary Radio Pulsars* (ASP Conference Series) **328** (2005) 25-31.

6.5.3 Thermodynamics of gravitational radiation

Let us begin with black-body radiation: where the spectral energy can be written as [37]

$$\mathcal{E}_{bb} = \bar{E}(\omega, T) \cdot \frac{\mathcal{V}}{\mathcal{V}_0}, \tag{6.28}$$

where $\bar{E}$ is some average energy that is a function of the angular speed, ω, and absolute temperature, T. To make the average oscillator energy a function of temperature, Planck introduced his random hypothesis that the field is made up of an incoherent superposition of modes in thermal equilibrium at a temperature T. The average oscillator energy is multiplied by the ratio of the volume of phase space occupied by the system, $\mathcal{V} = p^3 V$ to the minimum volume of phase space permitted by Heisenberg's uncertainty principle, $\mathcal{V}_0 = \hbar^3$. The momenta of the Planck oscillators are $p = \hbar\omega/c$, and V is the physical volume of the radiation cavity. Consequently, (6.28) can be written as a third-order aberration:

$$\mathcal{E}_{bb} = \bar{E}(\omega, T) \left(\frac{a\omega}{c} \right)^3. \tag{6.29}$$

Aberration is due to the different propagation speeds of the Planck oscillators in the walls of the radiation cavity in comparison to the speed of light, $\hbar/c \cdot a/h$, where a is the radius of the cavity. Once Planck found the correct weighting factor, an integration over all frequencies reproduced Stefan's T^4 law. That the energy turned out to be a power of the temperature greater than 1 meant some of the energy was being channeled into degrees of freedom (DOF) other than translational ones such as the creation of new quanta of energy.

For Planck, the average energy was $\bar{E} = n(\omega/T) \cdot \hbar\omega$, where $n(\omega/T)$ was Planck's distribution which had to be a function only on the ratio ω/T in order to satisfy Wien's displacement law. In order to derive Stefan's law, he had to consider the energy in each interval, $d\mathcal{V}/\mathcal{V}_0$, and integrate over all frequencies.

---

[37] Actually, was Planck did was to consider the spectral density in each frequency interval. This was necessary in order that when he introduced the weighting factor, which was a function of temperature, he would recover Stefan's T^4, which was experimentally verified. Here, we do not need to specify his weighting factor; the interested reader can find the full, historical, account of Planck's eloquent derivation in, B H Lavenda, *Statistical Physics: A Probabilistic Approach* (Dover reprint, 2016) Ch 2.

Rather, we shall set the average energy $\bar{E} = GM/a$, and not worry about the thermal factor for the moment. We can evaluate the numerator with the aid of Kepler's III, and when we do we get $(a\omega)^5/c^3$. Multiplying numerator and denominator by c^2, we get:

$$\mathcal{E}_{gw} = Mc^2 \left(\frac{a\omega}{c}\right)^5,$$ (6.30)

Multiplying both sides by ω gives the luminosity:

$$\mathcal{L}_{gw} = Mc^2 \left(\frac{a\omega}{c}\right)^5 \omega,$$ (6.31)

which has the same dependencies on angular speed and the speed of light as (6.26).

In comparison with the expression found for the quadrupole moment, (6.27), (6.31) is off by a factor of the square in the relative velocity. The rest mass, (6.22), is the same as that used in the Landau & Lifshitz treatment of GWs given above. Taken at face value one would expect that point particles in a 'cloud of dust' would follow geodesics, and not be candidates for emitting gravitational radiation.

Rather, the fundamental process is 5th-order, and because of the nature of the average energy gives a 7th-order one. This implies a DOS of the form:

$$\frac{\mathcal{V}}{\mathcal{V}_0} = \frac{p^5 V^{5/3}}{\hbar^5} = \left(\frac{a\omega}{c}\right)^5.$$ (6.32)

However, this looks suspicious since there is no radiation cavity that has a volume of $V^{5/3}$.

Moreover, if we now evaluate (6.31) using Planck's function, we come out with a T^5 law for radiation. This is because the 'average energy,' Mc^2, does not introduce an addition frequency as in Planck's case. In Planck's case, the exponent of the temperature less one represented the number of half -DOF.

In contrast, GR, would 'predict' a Stefan's law of T^7. For a photon gas, we have to double the 3 and get 6 DOF. Throwing out 4 because of the invariancy under spacetime translations of Maxwell's equations leave 2 which show up in the 2 independent states of polarization of the photon.

Whereas, for (6.31), the temperature exponent, T^5, indicates that there are 10 DOF. In addition to throwing out the 4 found in electromagnetism, this time supposedly because of the invariance of the Einstein field equations under coordinate transformations, another 4 should be discarded because of the Bianchi identities and again we come out with 2 independent states of polarization.

But, a T^7 law envisages 14 DOF. Getting rid of 8 for the reasons stated above would predict 6 DOF. This does not correspond to the helicity states of graviton . So something is amiss with a luminosity given by (6.27), or the Einstein equations themselves which allow getting rid of 4 DOF through the Bianchi identities.

A Stefan law of T^2 corresponds to first-order aberration, and a 1-dimensional system. For an ideal gas, the temperature is proportional to the mean kinetic energy, or the internal energy density, u. For degenerate systems, the temperature will only grow as $u^{1/(\eta+1)}$ because energy is being fed into other DOF. [38]

If we had accepted (6.26) as the sacred scriptures, we would be hard pressed to explain a radiation cavity having a volume of $V^{5/3}$. We would like to think that radiation cavities are determined by a 3-dimensional volume, which can also be the interior of a star.

In the above demonstration there is no problem in doing so since what we are dealing with is essentially black-body radiation with a different form of the average energy. This means that GWs are actually closer to EM waves than one would superficially expect.

Said slightly differently,

> electromagnetic waves have an intrinsic duality: they are necessarily also gravitational waves,

a sentiment echoed by Cooperstock. [39] We would even go further and claim that grav-

---

[38] B H Lavenda, *Thermodynamics of Extremes* (Horwood, Chichester, 1995) §II3.6.

[39] F I Cooperstock, "The essence of gravitational waves and energy," dated 26/03/2015.

ity has no propagational capacities in its own right, but, rather, employs electromagnetic waves for its transmission by modifying the index of refraction through which they propagate. But, even this is not wholly satisfactory since it would not explain the lack of shielding of GWs.

6.5.4 A critique of LIGO's analysis of GWs

There is nothing in Newtonian theory to tells us the speed at which forces propagate. To that we have to look at celestial secular perturbation theory.

If V is the propagational speed of the gravitational force, and a_0 the initial semi-major axis at time t_0 of an orbiting body, with μ as the standard gravitation parameter, then the decrease in the orbit later time, t, will be given by: [40]

$$a^2(t) = a_0^2 - 4\frac{\mu}{V}(t - t_0),$$ (6.33)

where V is the speed of propagation of the gravitational force. Differentiating (6.33) in time and equating it to the same term in the time derivative of $E/M = -\mu/a$ give:

$$\dot{a} = 2\mu/Va = (2a^2/M\mu)\dot{E}.$$

Rearranging, introducing Kepler's III, and solving for $\dot{E}$ result in:

$$\dot{E}/M = V^2 \left(\frac{wa}{V}\right)^3 \omega.$$ (6.34)

To get the rest energy, it behooves us to set $V = c$, the speed of light. Expression (6.34) is a third-order aberration, and arises because we introduced a finite propagation speed.

If we retain the speed, V, in (6.33), and equate $\dot{a}/a = (2/3)\dot{P}/P$, where $\dot{P} = (a\omega/c)^3$, we find

$$\frac{\mu}{Va^2} = \frac{1}{3}\left(\frac{a\omega}{c}\right)^3.$$

Then, availing ourselves of Kepler's III, we come out with:

$$\frac{c}{V} = \left(\frac{a\omega}{c}\right)^2.$$ (6.35)

---

[40]Danby, *op. cit.*

Not insisting on the value of the numerical coefficient in (6.35) shows that if $a\omega < c$, then $V > c$. Having injected this solely out of curiosity, we will from now take $V = c$.

On reading almost any tutorial for students on binary mergers, we are startled by two equations that seem to pop out from nowhere: [41]

$$P(t) = \left(P_0 - \frac{8}{3} \mathsf{M} t \right)^{3/8}, \tag{6.36}$$

where, dropping numerical factors, $\mathsf{M} = (G\mathcal{M})^{5/3}/c^5$ and

$$\mathcal{M}^{5/3} = c^5/G \frac{\dot{\omega}}{\omega^{11/3}}, \tag{6.37}$$

where $\mathcal{M}$ is commonly referred to as the 'chirp' mass. The term 'chirp' supposedly comes from the decreasing period and increasing amplitude of the gravitational wave signal as the stars approach each other and coalesce. We will not need it, but will comment on it later.

The rate of decrease in period, (6.36), is superseded by the generalization of the secular perturbation:

$$a^{\eta}(t) = a_0^{\eta} - \frac{\mu^{\eta-1}}{c^{\lambda}}(t - t_0), \tag{6.38}$$

where η and λ are two exponents. For $\eta = 2$, $\lambda = 1$, for $\eta = 3$, $\lambda = 3$, and for $\eta = 4$, $\lambda = 5$.

In the later case, differentiating:

$$a^4(t) = a_0^4 - \frac{\mathsf{M}^3}{c^5}(t - t_0), \tag{6.39}$$

and equating it to the expression containing $\dot{E}$, we obtain:

$$\dot{E} = Mc^2 \left(\frac{a\omega}{c} \right)^7 \omega, \tag{6.40}$$

where we have used Kepler's III for the chirp mass, $\mathcal{M}$. Expression (6.40) can rightly be considered as a one-step derivation of the GR luminosity (6.27).

---

[41]Teachers Guide, "Hands-on gravitational wave astronomy: Extraction of astrophysical information from simulated signals".

Equation (6.39) will suffice for our discussion of the coalescence of the two stars in the binary, without any reference to increases in the orbital frequency which cannot occur without simultaneous decreases in the semi-major axis that is dictated by Kepler's III.

Setting $a(t) = 0$ in (6.39), and assuming an initial centripetal velocity, $v_0 = \sqrt{M/a_0}$ where $v_0 = 2\pi a_0/P_0$, where $P_0 = 1.14 \times 10^{-4}$ yr, or 8 hours, is the initial orbital period. Introducing these values into the condition that the semi-major axis vanishes:

$$\left(\frac{a_0\omega_0}{c}\right)^5 = \frac{T_0}{\Delta t} = \frac{1.14 \times 10^{-4}}{108 \times 10^8} \sim 10^{-12}, \tag{6.41}$$

where we used the merger time of 108 Myr for PSR B1913+16. [42] The relative velocity on the left-hand side of (6.41) is of the order of 10^{-4}. This raised to the 5th power is 8 orders of magnitude smaller than the right-hand side. Rather, if the exponent is 3 we obtain a match, confirming the values in Fig (6.5).

6.5.5 The chirp mass

We are told that: [43]

> Like the radial velocity curve, the period (or frequency) of the LIGO signal is directly related to the orbital period (or frequency) of the binary system. In the LIGO case, however, the period of the detected gravitational wave is *half* that of the orbital period, so the wave frequency is *twice* that of the orbit.

If GWs have anything at all to do with their cousins, EM waves, then a body should emit a spectrum of them over a definite frequency range, like in the Wien and Rayleigh-Jeans limits of EM waves. Emitting a single monochromatic wave at exactly the frequency of the orbital period would be the ultimate free lunch.

If changes in gravitational fields are GWs then a whole spectrum of them should be produced not only by a merging binary system, but, simply by a man standing on Earth and holding a heavy club in his hand. [44]

---

[42]K A Postnov & L R Yungelson, "The evolution of compact binaries," *Living Rev Relativ* **9** (2006) 6, Table 3.

[43]physicsfromplanetearth.wordpress.com/2016/05/23/gravitational-radiation-2-the-chirp-heard-round-the-world/

[44]J L Synge, *Relativity* (North-Holland, Amsterdam, 1960) Ch IX

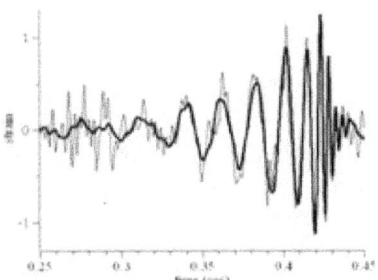

Figure 6.6: The graph of the gravitational wave 'strain' data, magnified some 10^{21} times, from the LIGO $GW150914$ event versus time in sec. The thin curve is the 'raw' data from the interferometer in Hanford WA, while the thick curve is the waveform according to NR. See: B P Abbott *et. al.* "Observation of gravitational waves from a binary black hole merger," *Phys Rev Lett* **116** (2016) 061102-1-15. Any model that employs Kepler's III has as $a(t) \to 0$, $\omega(t) \to \infty$, in such a way that $a^3(t)\omega^2(t) =$ const., and not to some finite value.

At first the club hangs down toward the ground, but, at a certain moment the man raises it quickly over his head. Any theory of gravitation recognizes that the club produces a gravitational field, however minute it may be, and that the action of the man changes that field, not only in his neighborhood, but throughout the whole universe . . . the change in the gravitational field of the moving club travels into space at the speed of light. And we would call this moving disturbance, a *gravitational wave*.

Everything and anything supposedly produces a GW so if it is a propagating disturbance what restricts it to a monochromatic wave with a frequency the same identical one as the orbiting binary? And the assumption that only massive binaries would produce GWs of sufficient magnitude to be observed is offset by their distance. Recall, this is what allows the Einstein equations to be linearized about the flat Minkowski metric in the first place.

The detected signal, shown in Fig (6.6), shows that the signal increases in frequency from 35 Hz to 150 Hz in the time interval 0.250 to 0.425 s immediately before the two bodies coalesce.

Equation (6.37) is integrated to give:

$$GM^{5/3}(t_2 - t_1)/c^5 = -A\left(\omega^{-8/3}(t_2) - \omega^{-8/3}(t_1)\right),\qquad(6.42)$$

where A is a numerical constant. Equation (6.42) is a mere re-write of (6.39) on the strength of Kepler's III. When the above numbers are plugged in the chirp mass $\mathcal{M}$ is determined, which must be a constant throughout, again, on the strength of Kepler's III.

However, if Kepler's law applies at the point of merger, $a(t_2) = 0$, and the corresponding frequency $\omega = \infty$!

It is not our intention to provide a critical analysis of the LIGO interferometer. But, one thing is clear that like an EM wave which creates an EM field as it passes, so, too, must a GW wave create a local gravitational field as it passes.

Even according to LIGO's own admission, when it considers the transient GW in the interferometer: [45]

> Since the lengths of the arms are constantly changing as the gravitational wave passes, so too does the amount of out-of-sync-ness of the light waves!
> While this sounds like a mess, it's not In fact, over the course of the gravitational wave's passage, the resulting interference pattern changes in-step with the changing lengths of the arms .. the actual wavelength changes of the beams of light have no bearing on the much more important interference pattern. The effect of the length changes in the arms far outweigh any change in the wavelength of the laser, so we virtually ignore it altogether.

This wishful thinking for it implies that light, and the speed that it propagates, are unaffected by a gravitational field. This is in flagrant negation to the whole theme of this book, not to say that it negates what Einstein himself claimed!

---

[45] www.igo.caltech.edu/page/faq

We have constantly emphasized that as light passes through a gravitational field, it will feel an index of refraction that will affect its speed of propagation. This is more than enough to modify the interference pattern that "LIGO scientists can look at the changing interference pattern and decipher exactly how the arms of the interferometer itself must have changed over time to create the patterns that they observe."

6.5.6 NR versus GR

The spectra LIGO compared their signals to are derived from NR. In this section we show that NR, instead of supporting GR, in fact, contradicts it.

NR is based on a $(3 + 1)$ decomposition of Einstein's field equations. It is known that *there is no preferred time-slicing in GR, and, consequently no unique definition of the 'instantaneous' direction.*

The standard form of the metric that NR employs is:

$$ds^2 = \alpha^2 dt^2 + \gamma_{jk}(dx^j + \beta^j dt)(dx^k + \beta^k dt),$$

in 'natural' units where $c = G = 1$, and the indices i, j run over the all space.

Some of Einstein's equations do not contain time derivatives; they are used as constraints. The two constraint equations are:

$$R + K^2 - K_{ij}K^{ij} = 16\pi\rho,$$

and

$$D_j(K^{ij} - \gamma^{ij}K) = 8\pi S^i,$$

where the Christoffel symbol, $K_{ij} = \sqrt{-g^{00}}\Gamma^0_{ij}$ is referred to as extrinsic curvature of a hypersurface since it gives the variation of normal vectors to the hypersurface from point to point. The covariant derivative with respect to the γ_{ij} is symbolized by D_j. The density, ρ, and flux, S^i, are expressed in terms of an energy-matter tensor T^{ij}, representing some fictitious hydrodynamic fluid which supposedly obey polytropic equations. Such equations assert a relation between changes in density and pressure, as a function of radius, independent of the temperature. They are therefore not considered as equation of state.

The evolution equation for the space metric is:

$$\frac{\partial \gamma_{ij}}{\partial t} = -2\alpha K_{ij} + D_i \beta_j + D_j \beta_i,$$

and a more complicated one for the evolution of the extrinsic curvature, which need not concern us here. For the solution of the Cauchy problem, it is necessary to specify the initial values of γ_{ij} and its time derivative at the initial time $t = t_0$, at all points in space.

Invariancy with respect to time and space translations is supposedly taken up by α, the 'lapse' function, and the β^i, are called the 'shift' vector. The lapse function gives the freedom to choose the space-like hypersurfaces, while the shift vector embodies the liberty to label the spatial coordinates on any hypersurface in any way we like.

However, this is not a reflection of Hilbert's criterion [46] that *all physical results which are deducible from Einstein's equations must be independent of the coordinate system in which the solution is expressed*. Only when $\alpha = 1$, and all the β^j vanish does the metric reduce to its Gaussian-normal form. [47] In other words, it violates Einstein's desire that "the results of general relativity," reflect the "general covariance of the laws of nature."

Another blemish of NR is its employment of the gravitation pseudo-tensor. The often employed ADM mass (Arnowitt-Deser-Misner) is based on this pseudo-tensor, which destroys, *a priori*, its validity.

For this it is sufficient to repeat the words of Levi-Civita [48]

> differential invariants of the first-order which are intrinsic, exclusively formed from the coefficients of the metric and their first derivatives, do not exist. This is enough to render inadmissible the form of Einstein's gravitational tensor.

He then goes on to tell us that Einstein was already uneasy about this prospect, for

---

[46]D Hilbert, *Gesammelte Abhandlungen*, Dritter Band (Springer, Berlin, 1935) 258.
[47]L D Landau & E M Lifshitz, *Theory of Fields* (Pergamon, Oxford, 1975).
[48]T Levi-Civita, "On the analytic expression that must be given to the gravitational tensor in Einstein's theory," *Rendi Reale Accad Lincei* **26** (1917) 381.

"after having outlined with genial simplicity the theory of gravitational waves, he was led to the unacceptable result that *spontaneous* waves should as a rule give rise to a dispersion in energy through radiation."

Levi-Civita also provides a panacea to Einstein's predicament, and that is to have the pseudo-tensor vanish along with the energy-stress tensor, so that would entail "a total lack of stresses, of energy flow, and also of a simple localization of energy." Truly, a gruesome prospect.

However, the gravitational pseudo-tensor cannot be appended to the Einstein field equations, and the fact that the Einstein tensor, G and the energy-stress tensor, T, each satisfies a zero 4-divergence condition. If these represent conservation conditions, it means that there is no 'communication between', or 'transfer from', the worlds of geometry, embodied in G, and energy-stress and matter, T.

NR also predicts the "inspiraling and merger of binary black holes." Yet, *there does not exist a solution to Einstein's field equations for two bodies, and certainly not their singularities.* In the words of Levi-Civita, [49] "Even the two body problem, solved long ago by Newton, has very little chance of being successfully solved in terms of General Relativity." In view of this how can NR come to the conclusion that "the end product of a black hole binary coalescence is a Kerr black hole, which is fully described by its mass and spin"?

Claims have been made that NR religiously respects event horizons, and that "Einstein equations [are] relevant only for densely, compact objects: black holes and neutron stars." [50] What then is the force that would attract the two BHs to coalesce if nothing can escape the event horizon? How can a numerical procedure shore up deficiencies in a theory that neither makes such predictions nor pretends to describe such phenomena?

Einstein [51]himself was the first to admit this:

---

[49]T Levi-Civita, *The n-Body Problem in General Relativity* (Reidel, Dordrecht, 1964) p VIII.

[50]E Schnetter, "Introduction to numerical relativity," talk given at Caltech on 26/07/2013.

[51]A Einstein, *Out of My Later Years* (Philosophical Library, New York, 1950) pp 94-95.

One thing, however, is certain: if a field theory results in a representation of corpuscles free of singularities, then the behavior of these corpuscles with time is determined solely by the differential equations of the field.

Recognizing the problem of $r < 2m$, Einstein attempts to construct a 'bridge' connecting two shells neighboring on a hypersurface $r = 2m$ which would allow him to conclude that

the existence of such a bridge between the two shells in the finite realm corresponds to the existence of a material neutral particle which is described in a manner free from singularities.

In the interim period, the equations haven't changed, but their interpretation has!

And there is the fact that all solutions to Einstein's equations are geodesic paths which do not cross one another. They can either bunch up or spread out depending upon which non-Euclidean geometry is being considered, but their behavior is not caused by forces, even tidal ones. [52] That would destroy the whole idea that GR is founded on. How can an innocuous 'cloud of dust' ever generate a GW? [53] or two colliding GWs create a singularity? [54] Such phenomena must be accounted for in the energy-stress tensor, albeit nothing is known beyond an ideal fluid.

We have remarked on how accelerations are destitute of any physical meaning in GR. The motion of particles on a Minkowski background are geodesics: their 4-vector u^i satisfy $u^i \nabla u_i = 0$ meaning that the motion is both rectilinear and uniform. As Weyl [55] remarks "From the standpoint of Einstein's theory this is as it should be, because the gravitational force arises only when one continues the approximation beyond the linear stage." So we must resort to perturbations in order to derive a gravitational force!

---

[52]H Ohanian & R Ruffini, *op. cit.*

[53]A Loinger & T Marscio, "Remarks on numerical relativity, geodesic motions, binary neutron star evolution," arXiv:1211.6152.

[54]D'Inverno, *op. cit.* §208.

[55]H Weyl, "How far can one get with a linear field theory of gravitation in flat space-time?" *Am J Math* **66** (1944) 591-604.

And even if that is not sufficient, Weyl [56] comments on the notion of motion in GR that "relative motion of several bodies has, as the postulate of general relativity shows, no more foundation than the concept of absolute motion of a single body." GR always guarantees that there is a coordinate transformation that will allow us to pass from bodies in motion to a co-moving one in which bodies are at rest with respect to that frame. This not only rids us of magnetic fields, but, more important in the GR context, that bodies cannot generate GWs.

As we have mentioned before, a linearized Riemann tensor with an indefinite metric would give rise to a wave equation. Yet, Einstein's concept of the universe was, at least initially, eternal where heavenly bodies followed geodesic paths.

And finally, there is the incomprehensible desire to extend the Schwarzschild outer solution *beyond* its domain of validity. In his *Absolute Differential Calculus*, Levi-Civita gives rigorous meaning to a metric, like the Schwarzschild one, as being spherically symmetric in a curved space manifold. In it, he shows that at great distances from the attractive mass, the coordinate r "becomes identical with the length of the radius vector drawn from the centre of symmetry; further, the expression: [57]

$$-\frac{1}{2}c^2 V^2 = -\frac{1}{2}c^2 + \frac{1}{2}c^2 \frac{\alpha}{r}$$

represents the potential of the field." This implies $r > \alpha$, and *under no circumstances can the roles of space and time be inverted, giving $r < \alpha$.*

Then what about the event horizon? Late in life Hawking had misgivings about the event horizon when confronted with the concocted 'information paradox' where information can be lost forever when swallowed up by a BH. In his opinion, BHs turned into grey instead of being black. But where in the history of science does a personal opinion determine a physical law of nature? It happens solely when there is no experimental verification. The same also applies to his relation between surface gravity and absolute temperature, and to the Bekenstein-Hawking expression for entropy of a BH, based on an erroneous analogy with the area of the event horizon. [58] To add insult to injury it is claimed that there is nothing to worry about because the Schwarzshild

---

[56]H Weyl, *Space, Time, Matter* (Dover, New York, 1952).

[57]T. Levi-Civita, *Absolute Differential Calculus* (Blackie, London, 1926) p 423.

[58]B Lavenda, *Where Physics Went Wrong* (World Scientific, Singapore, 2015 §6.3.

radius "is not a physical singularity [because] it arises from an inappropriate choice of coordinates and can be eliminated by a change of coordinates." [59]

Coordinate changes can do many strange and unphysical things as Eddington realized. We have shown that they cannot be applied to non-Euclidean geometries without changing their characteristics. As Eddington contends, the ideal propagation of coordinate changes occurs at the *speed of thought* giving rises to a periodic dependent metric whose waves would propagate just as fast.

6.5.7 Michelson interferometer inundated by GWs

The question has arose occasionally on how can interferometers be used as GW detectors when light waves will also be affected by the passage of a GW wave? If the answer is yes, then the wavelength will also change leading to a change in the speed of propagation of the laser beam. In this section, we shall use a simplified model [60] that was designed to show that the phase of laser beams are modulated by GWs.

It's been claimed that LIGO is the most precise ruler that exists. In Fig (6.6), the units are those of relative change in length, or what LIGO refers to as 'strain'. The values are in one part in a thousand trillion trillion. With 4 km arms, shown in Fig (6.7

the actual change in length is 4×10^{-18} m, or about $1/1000^{th}$ the size of a proton.

The change in distance is measured by a difference in phase between laser beams sent out and reflected back along each of the two orthogonal arms. When combined at the detector they produce an interference pattern if the beams are out of phase.

This supposedly is what happens when a GW passes through the interferometer. The supposition is that one arm will be stretched while the other is squeezed. A laser having a wavelength of $1\mu m$ or about 10^{-6} m a variation of 10^{-8} m abounds to 10^{-12} wavelengths. But, the GW will also affect the wavelength of the beam so there should be no change in the number of wavelengths.

---

[59]Ohanian & Ruffini, *op. cit.* p 326.

[60]P R Saulson, "If light waves are stretched by gravitational waves, how can use light as a ruler to detect gravitational waves?" *Am J Phys* **65** (1997) 501-505.

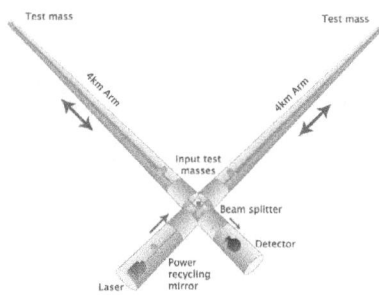

Now, the basic 'trick' that LIGO employs is that the speed of light should remain constant, so if one arms gets longer, light should take a longer time to traverse the arm and when it returns to the detector it should produce a minute change in its phase with respect to the phase of the beam in the other arm.

We will employ a simple metric often used to demonstrate this effect. Yet, we will find that the velocity of light is not constant, but, rather a function of the amplitude of the GW. This can be interpreted as a change in the index of refraction when light passes through a medium with a different index of refraction. If the index of refraction is n, the speed of light is c/n. Let us consider this in slightly greater detail.

This is a classic example of the use of a Michelson interferometer. A glass plate of thickness δ and index of refraction n is inserted normal to the path of one of the two interfering beams. Whereas the optical path length of the beam through the plate is $n\delta$, the optical path length through an equal thickness containing air only is just δ. Therefore, there will be an increase in optical path length of $(n-1)\delta$ due to the insertion of the glass plate. Since the beam is send back and forth, a factor of two must be inserted. Then, if N is the number of fringes displaced when the glass plate is

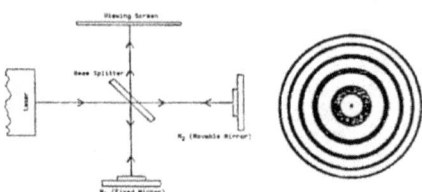

Figure 6.8: A Michelson interferometer with movable mirror, M_2.   On the right is
the interference pattern with the dark spot in the center.   Taken from
mini.physics.sunysb.edu/ mdawber/phy252:michelson

inserted, it follows that $N\lambda = (n-1)\delta$.

The mirrors are adjusted to obtain circular fringes with a single dark spot at the center,
as in Fig (6.8). Subsequently, when the plate is introduced in one of the arms, the fringes
are displaced. The movable M_2 mirror in that arm is moved a distance Δ closer until
the single central dark spot again results. Both the number of fringes N that have
disappeared and the distance that the mirror is moved, Δ, are registered. Since the
glass plate has increased the optical path length by $2(n-1)\delta$, and the movement of
the mirror decreased it by 2Δ, the two effects must be equal,

$$N\lambda = 2d = 2(n-1)\delta,$$

and from which the index of refraction, n, is:

$$n = 1 + \frac{N\lambda}{2\delta}. \tag{6.43}$$

Knowing the thickness of the plate, the number of fringes that have disappeared, and
the wavelength of light, the index of refraction can be computed. But, if the wavelength
doesn't change, and surely the frequency of the light doesn't, how can their be a change
in the speed of light? And yet, we know that when light passes through a medium of
an index of refraction, $n \neq 1$, the speed does change.

Introducing the wave number, $k = 2\pi/\lambda$, Eq (6.43) can be written as

$$k' - k = \pi N/\delta, \qquad (6.44)$$

where the wave number, $k' = nk$. Only when $n = 1$, does the effect disappear and the speed of light returns to its vacuum value, c. [61]

The speed of light passing through the arm of the interferometer has been decreased by an amount c/n because of the insertion of the glass plate. We claim that the passage of the GW should have the same effect as the insertion of the glass plate.

Parenthetically, it is interesting to note that (6.43) can be used to determine all the 'classical' tests of GR. The relevant index of refraction is:

$$n = 1/(1 - \alpha/r) \approx= 1 + \frac{\alpha}{r},$$

which is that obtained from the Schwarzschild metric. Introducing it into (6.43) results in:

$$\frac{N\lambda}{\delta} = 2\frac{\alpha}{r}.$$

The right-hand side gives the deflection of light when r is set equal to the radius of the Sun, the perihelion shift when r is the semi-major axis of the elliptical orbit, and for the red-shift r is the object radius. So we have a good idea of the magnitude of $N\lambda/\delta$ in the case of gravitational phenomena.

Now suppose that a GW is propagating along the z-axis and has its polarizations aligned with the x- and y-axes. The metric for such an arrangement is

$$c^2 d\tau^2 = c^2 dt^2 - [1 + h(t)]dx^2 - [1 - h(t)]dy^2, \qquad (6.45)$$

where the wave amplitude $h(t)$ of the GW is a sole function of time. The metric has been so designed that whatever occurs along the x-axis, i.e. a shrinking or expansion, will occur with equal and opposite changes along the y-axis. Supposedly, sending light

---

[61] Interferometers present several interesting paradoxes. We have calculated the total time, (5.67), it takes to complete a round-trip with a speed on the outward journey, $c + v$, and that on the return journey, $c - v$, where v is the velocity relative to the source. But, if c is the limiting velocity there should be no 'contraction' of the arm length in the direction of motion, and no longitudinal Doppler effect [cf. L Essen, *The Special Theory of Relativity: A Critical Analysis* (Clarendon Press, Oxford, 1971).]

signals back and forth *at the speed* c will show up as a displacement of fringes when the two light waves are recombined.

Light rays follow null geodesics, $d\tau = 0$, so we have to indicate their progression by an affine parameter λ. Using a dot to indicate the derivative with respect to this parameter, the Lagrangian can be read off the metric (6.45) as:

$$\mathcal{L} = \frac{1}{2}\left\{ c^2\dot{t}^2 - \dot{r}^2 - r^2\dot{\varphi}^2 + 2h(t)r\sin 2\varphi\,\dot{r}\dot{\varphi} \right\}, \qquad (6.46)$$

upon transforming to plane polar coordinates, (r, φ).

The usual Euler-Lagrange equations give the geodesics. There are two cyclic coordinates, t and φ, so we would expect two first integrals of the motion. Rather, on account of the GW wave, there will be no energy integral because

$$c^2\frac{d}{d\lambda}\dot{t} = \dot{h}r\sin 2\varphi\dot{r}\dot{\varphi} \neq 0,$$

while the angular momentum will be given by

$$L = r^2\dot{\varphi} - hr\sin 2\varphi\dot{r}. \qquad (6.47)$$

The geodesic equation for the radial coordinate,

$$a_{rad} = \ddot{r} - r\dot{\varphi}^2 = \dot{h}\dot{\varphi}r\sin 2\varphi + 2ht\cos 2\varphi\dot{\varphi}^2 + hr\sin 2\varphi\ddot{\varphi}$$

showing that the radial acceleration is coupled to a component of the tangential acceleration, $a_{trans} = r\ddot{\varphi} - 2\dot{r}\dot{\varphi}$.

On the strength of the vanishing of the Lagrangian, (6.46), the angular momentum may be written as

$$L^2/r^2 = [c\dot{t} - \sqrt{1 - h^2\sin^2 2\varphi}\,\dot{r}][c\dot{t} + \sqrt{1 - h^2\sin^2 2\varphi}\,\dot{r}]. \qquad (6.48)$$

Equation (6.48) shows that light propagates along the arm in both directions with a speed:

$$c/\sqrt{1 - h^2\sin^2 2\varphi},$$

and not at speed c. As in the above example of a glass plate, the wavelength has stayed the same but light will travel at a slower speed because it is traversing a medium with a larger index of refraction.

Before the arrival of a GW, the time taken for a pulse to traverse the length L of either of the arms and return is $2L/c$. Upon the arrival of a GW, one arm will be shortened and the other lengthened. The supposition that there will be a difference in arrival time of the two pulses, $(2L/c)h_0$, where h_0 is the assumed constant value of the GW amplitude hinges on the fact that the speed of light is c. This does not constitute a way of measuring the strength of a GW.

While it is true that the speeds of reference frames in GR are completely arbitrary, the speed of light is affect by a GW, if the model has any correspondence with reality.

In conclusion, we may say that GR has the same limitations as GO. This is why Landau & Lifshitz [62] could so easily obtain the path trajectory of a particle moving in a centrally symmetric gravitational field by employing the relativistic Hamilton-Jacobi equation, which the basic equation of geometrical optics [63] without resorting to Einstein's field equations. *The phenomena predicted by NR lie outside the domain of validity of GR.*

6.6 Problems with magnetic fields

It has been realized that a magnetic field from a Schumann resonance can affect the LIGO interferometer in the exact same way as a GW. [64] An ambient magnetic field can affect the interferometer because tiny magnets are used to control the position of the mirrors. Moreover, it can induce a current in a wire that is part of the detector. The current carrying the GW signal passes through the electromagnetics which restore the mirrors back to their original position. Hence, the presence of ambient magnetic fields can have devastating effects on the credibility of LIGO's detection of GWs.

---

[62]Landau & Lifshitz, *op. cit.* §101.

[63]L D Landau & E M Lifshitz, *Fluid Mechanics* (Pergamon, Oxford, 1959) §66, specifically Eq (66.3).

[64]C Daniel & R Schofield, "Magnetism and Advanced LIGO".

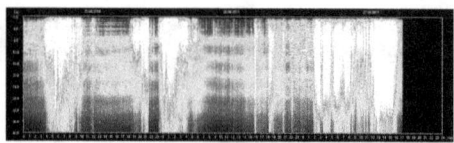

Figure 6.9: Schumann resonances (in white) taken from the Russian Space Observing System in Tomsk.

A natural hohlraum, or radiation cavity, looms above the earth extending from some 30 miles above the earth to 600 at the edge of space. In 1954, Schumann detected resonances at the fundamental frequency of 7.83 Hz which is a global electromagnetic resonance that is generated by lightning discharges in the ionosphere. Lightning can be considered as a sudden flow of charges between oppositely-charged surfaces. The accelerating charges generate electromagnetic waves propagate spherically, and resonate in the hohlraum produced by the earth's surface and the ionosphere.

Schumann resonances consist of distinct spikes that occur at extremely low frequencies (ELF). They are spaced at around 6.5 Hz apart: the fundamental at 7.83 Hz, and higher harmonics at 14.3 Hz, 20.8 Hz progressing until 59.8 Hz. However, these higher harmonics seem to be the exception rather than the rule.

They were originally predicted by FitzGerald in 1893 who observed that the upper atmosphere is a fairly good conductor He estimated that the electromagnetic oscillations would be at the lowest mode of the Schumann resonances. Almost a decade later, Heaviside alluded to the fact that the hohlraum can trap ELF electromagnetic waves.

Interest in Schumann resonances have come from an entirely unexpected group concerned with the 'science of interconnectivity,' which associates the resonances with human brainwave activity. Be that as it may, monitors on the Russian Space Observatory recorded a sudden spike in activity around 8.5 Hz in June of 2014. The white areas in Fig (6.9) are individual spikes in the Schumann resonance between $12 - 16.5$ Hz.

The fundamental resonant frequency has been steadily increasing since June 2014. On January 31, 2017, a five-fold increased was recorded, from its baseline of 7.83 Hz, where it surpassed 36 Hz, which is right smack in middle of the range of the 'observed' GWs. New events, which the LIGO-Virgo collaboration attributes to binary black hole mergers, were designated by them $GW\,170729$, $GW\,170809$, $GW\,170818$, and $GW\,170823$, referring to the dates when the putative observations were made.

Even if what LIGO is observing are not Schumann resonances, an ambient magnetic field cannot only move the mirrors by affecting the tiny magnets but it can moreover induce a current in the wire that supposedly the brightness in the laser light produces when it returns the mirrors back to their original positions. In time, this current supposedly encodes the GW signal. This raises the question of whether the quantum limits, set-down by Heisenberg's uncertainty principle have been violated.

6.7 Quantum non-demolition measurements

Beginning in 2024, the enhanced Advanced LIGO Plus detector will include the Heisenberg uncertainty principle to "improve the machine's ability to detect ripples in space-time." [65] So what has it been doing till now?

LIGO boasts of an astonishing precision of one part in one hundred billion. The measurements made on GWs emanating from two binary neutron stars are no bigger than one thousandth of a proton radius.

A new version of the detector would minimize quantum fluctuations that affect the laser light. The idea is to use a crystal with nonlinear optical properties that would increase one of the conjugate variables at the expense of the other. In this case the amplitude fluctuations would increase at the expense of the phase noise. This is supposedly accomplished by 'squeezing' the vacuum which has phase fluctuations smaller than the normal normal vacuum. It is the phase shift which what of interest to LIGO.

However, amplitude and phase shifts must both satisfy the conservation of energy. Recall that Einstein derived his coefficients of spontaneous and stimulated emission, and absorption using a 'dynamic equilibrium,' that was tantamount to energy conser-

---

[65] "Space-time Uncertain" - LIGO goes quantum," *The Daily Galaxy* posted 21/02/2019.

vation. However, Einstein failed to specify the phase difference between the incident and emitted electromagnetic field. Since it derived Planck's distribution, and has been found consistent with quantum mechanics, does the phase shift really matter? This would be analogous to the omission of spin in the Schrodinger equation.

However, only over very short times can energy conservation be violated according to Heisenberg's uncertainty principle. It certainly cannot be violated in a resonator cavity where a source of monochromatic light is emitted into a resonator and reflected back and forth between mirrors whose only loses are with the interaction of its mirrors. But this is like the boring out of a tiny hole in the hohlraum which is big enough so as not to disturb the equilibrium between the various modes of the electromagnetic spectrum. Although there is no overall consensus to the phase differences occurring in emission and absorption processes, [66] there can be an arbitrary phase shift accumulated over a round-trip only if the interferences between forward and back amplitudes compensate exactly. In Einstein's case, it is the vanishing of the interference term between the electric-field wave amplitudes that allowed him to neglect (unintentionally, of course) the phase change. So amplitude fluctuations, which make measuring lower frequency GWs more difficult, cannot simply be neglected.

Vacuum fluctuations that occur over a very short time scale can violate energy conservation. But spontaneous emission, which can be considered as a consequence of vacuum fluctuations, must obey energy conservation just like its counterpart, stimulated emission. [67] And it is precisely these processes, together with absorption, that are relevant to an optical cavity in which a beam with reflect a very large number of times without significant loss.

Moreover, there is the mechanical part of the interferometer which certainly comes under the protective wing of Heisenberg's uncertainty relation. Measuring such small displacements in the mirror make its momentum virtually unknown, according to Heisenberg's uncertainty principle.

And if these improvements are really necessary to obtain undisputed results, what has been done so far? Measurements of such tiny displacements of the mirror would make

---

[66]M Pollnau, "Phase aspect in photon emission and absorption," *Optica* 5 (2018) 465-473.
[67]Pollnau, *op. cit.*

Chapter 6. The electrodynamics of GWs
6.7. Quantum non-demolition
measurements

the momentum of the 40 kg mirrors virtually unknown. They would oscillate back and forth without ever coming to rest. If if that wouldn't be enough to dissuade using a Michelson interferometer for detecting GWs, the 4 km arms are no match for the tens of thousand of km wavelengths of such low frequency gravity waves.

Moreover, there is the problem that you cannot localize a body smaller than its wavelength. The LIGO interpretation of a GW travelling down to earth and stretching space along one arm while compressing it along the other arm at periodic intervals is untenable. The passing GW will 'see' a speck, and certainly that 'speck' cannot make out the properties of the wave like its polarization.

At any given instant, so LIGO claims, more space along one arm and less in the other will cause the laser beam to travel different distance so that they will recombine in different phases. The resulting brightness is used to produce a current that returns the test masses back to their original positions. However, the minute measurements of the displacement of the mirrors would make the their momenta completely unknown make it difficult for the current, or any other means, to return them to their initial positions.

Recognizing the problem at hand, Thorne writes in the forward to *Quantum Measurement* [68] belittling the textbooks on quantum mechanic written during the period $1940 - 1970$. by saying that they have "little respect or interest in the quantum theory of measurement." Surely, he has not read David Bohm's *Quantum Theory* published in 1949. There, he spells out in detail the quantum theory of measurement, and why Heisenberg's uncertainty principle has to be reckoned with in *all* quantum measurements.

After admitting that the act of "measurement has produced an irreversible and indeterminate change in the quantum object," the authors then go on to define a "nondemolition" quantum measurement. Their motivation was:

> In the 1970s, in connection with efforts to construct detectors for gravitational waves, it became necessary to invent methods for measuring macroscopic observables at levels of precision approaching and exceed-

---

[68]V B Braginsky & F Ya Khalili, *Quantum Measuremet* K S Thorne, ed. (Cambridge U P, Cambridge, 1992).

ing the standard quantum limits However, theoretical analyses of typical, traditional schemes of measurement showed that their precisions can never exceed the quantum limit, even in principle. The solution to this dilemma, it was recognized, was to use a nontraditional class of measurement schemes, carefully crafted to overcome the standard quantum limits. For these schemes was coined the term 'quantum non-demolition [QND] measurements.'

Their panacea is to take "long enough for the measurements [so] one can obtain any desired sensitivity." For instance, the minimum error in the measurement of energy is of the order $\hbar/\tau$. All that is necessary is to let time $\tau \to \infty$. This then defines a vacuous steady state in which the energy can be measured with unlimited precision but absolutely nothing can be said about the time needed to perform the measurement. And all types of unintentional interactions have certainly occurred within that time period.

They then go one to illustrate how such a QND measurement can be performed. Radiation pressure is considered in which it is necessary to consider an extremely small pressure, even at optical frequencies. To obtain such weak pressures it is necessary to "register the pressure produced by a few quanta."
However, the fluctuations will be so great that one cannot measure such a radiation pressure. Try measuring the "pressure" attributed to a single particle in a box! It is a well-known thermodynamic consequence that below a certain level, usually taken as as the validity of Stirling's approximation, that macroscopic measurements cannot be performed on such systems. [69]

---

[69]B H Lavenda, *Statistical Physics: A Probabilistic Approach* (Dover, New York 2015).

Epilogue

If anything, the discovery of GWs should be more alarming than reassuring. It is a clear case where the outcome was determined before the experiment was done! Yet, GR is certainly not the first theory to have predicted them.

Maxwell gave up trying to extend his field theory of EM to gravitation. For where do you find a situation where, when matter is widely separated, the forces are the least while the potential energy is greatest, whereas when the potential energy is least when the forces are greatest? This meant to Maxwell that he was dealing with a field with *negative* energy — something he found so repugnant that he gave the matter up entirely. But that did not carry over to his protégé, Oliver Heaviside.

Though his disciple, Oliver Heaviside, [70] did not, and went ahead to derive a theory of gravitation that was based on the non-instantaneous propagation of the gravitational force. At the end of his expose he admitted that it

> does not enlighten us in the least about the ultimate nature of gravitational energy. It serves, in fact to further illustrate the mystery. For it must be confessed that the exhaustion of potential energy from a universal medium is a very unintelligible and mysterious matter.

Heaviside modelled the gravitational acceleration $\vec{g}$, after an electric field. If it is the gradient of a scalar potential there is no further ado. But, citing Newton's letter to Bentley, he said that "it is as incredible now as it was in Newton's time that gravitative influence can be exerted without a medium," and that in that medium it propagates at a finite speed, v.

Heaviside considers $\vec{g} = -\nabla \phi$ as the condition that "the gravitational force is *exactly* dependent on the configuration of the matter." This he takes equivalent to $\nabla \wedge \vec{g} = 0$. So when this is not so the last relation must be invalid. And if it is invalidated, it should

---

[70]O Heaviside, "A gravitational and electromagnetic analogy," in *Electromagnetic Theory* Vol. I (The Electrician, London, 1898) pp 455-466.

be replaced by:

$$cv^2 \nabla \wedge \vec{g} = \dot{\vec{h}},$$ (6.49)

which introduces an *auxiliary* field, $\vec{h}$, where c is a constant.

If the auxiliary field, $\vec{g}$, is divergence free, its curl must be non-zero,

$$\nabla \wedge \vec{h} = -c\dot{g},$$ (6.50)

whose source term Heaviside modeled after Maxwell's displacement current, $-c\ddot{\vec{g}}$.
Then taking the curl of (6.49), and the time derivative (6.50), the auxiliary field can
be eliminated to obtain:

$$-v^2 \nabla \wedge (\nabla \wedge \vec{g}) = \ddot{\vec{g}}.$$ (6.51)

which is Heaviside's *transverse* wave equation since:

$$\nabla^2 = \nabla \mathrm{div} - \mathrm{curl}^2.$$ (6.52)

The vector identity (6.52) implies $\nabla \cdot \vec{g} = 0$. This motivates considering the analogy
with EM where $\vec{g}$ would be analogous to an electric field:

$$\vec{g} = -\nabla\phi - \frac{\partial \vec{A}}{\partial t},$$ (6.53)

which represents an "oscillating gravitational radiation field.[71] It should also

> be transverse to the wave propagation direction and to have an ampli-
> tude that falls off as $1/r$, the usual space dependence for the amplitude of
> [EM!] waves far from their sources.

Consequently, only the second term in (6.53) subsists so that "the gravitational radiation field $\vec{g}$ depends only on the components of the vector potential $\vec{A}$ that are
transverse to the observation direction:

$$\vec{g} = -\frac{\partial \vec{A}}{\partial t}.$$ (6.54)

---

[71]R C Hilbron, "Gravitational waves from rotating binaries without general relativity" a tutorial, 09/2017.

So that if $\nabla \cdot \vec{g}$, so, too, will $\nabla \cdot \vec{A} = 0$, which is referred to as the Coulomb gauge.

Is the analogy with EM waves relevant? EM waves are the only known waves that do not need a material medium to propagate in. Some media can support both transverse and longitudinal waves, such as ocean waves. Now if GWs are *mechanical* transverse waves they need a material medium, and propagate by means of vibrations that are perpendicular to the direction of propagation: the so-called 'ripples' of spacetime. This is like some sort of jello which robs the vacuum of its vacuous state. We have come full circuit by re-introducing the ether that Einstein say fit in 1905, but later repented having done so in 1920.

It is the auxiliary field, $\vec{h}$, that allows EM waves to propagate in "vacuum", and this has been borrowed by GR, and named the gravito-magnetic field. If it exists then the ripples on spacetime are completely superfluous so that GWs can propagate throughout the universe just like EM waves, and since they travel at the same speed, this would be the more logical choice. The two orthogonal vector fields ride piggy-back, and this is what allows the EM waves to propagate in a vacuum. The presence of a material medium tends to slow down their propagation by offering resistance to their propagation.

However, if they do need a material medium to propagate in, the polarizations of GWs must be characteristic of that medium rather than an intrinsic property of the waves themselves. And because they are considered as undulations in spacetime, they must show optical diffraction phenomena and aberration. Heaviside, too, was perplexed:

> The remarks of the Editor and of Prof Lodge on gravitational aberra-
> tion, lead me to point out now some of the consequences of the modified
> law when we assume that the ether is the working agent in gravitational
> effects, and that it propagates at speed v.

Heaviside considers how the gravitational force between the Sun and the Earth, f, is modified "when the Sun is in motion at speed u through the ether. The modified force law,

$$F = f \times \frac{1 - \beta^2}{(1 - \beta^2 \sin^2 \theta)^{3/2}}, \tag{6.55}$$

where $\beta = u/v$ is the relative velocity and θ is the angle between the line of motion and the radius vector between the Sun and the Earth. Except for the error that the term in the numerator $1 - \beta^2$ of (6.55) should be squared, the expression is identical to the result found by Ibison *et. al.* [72] found in §4.6, and pre-dates the latter by more than a century

Moreover, like IPL and Carlip, [73] Heaviside can't answer the objections of Lodge because (6.55) is a radial, and not a tangential, force.

What more can GR add? Eddington [74] addressed the problem, referring to two earlier papers by Einstein who investigated the propagation of GWs with the speed of light "due to changes in the distribution of matter," not mentioning that only accelerating masses are able create "the altered curvature of space-time," that resulted from gravitational radiation

Since the theory is tensorial, it admits three types of GWs: longitudinal-longitudinal, longitudinal-transverse, and transverse-transverse. [75] Eddington shows that only the latter "are propagated with the speed of light *in all systems of co-ordinates.*" Einstein found that the first two categories of waves "convey no energy." This seems a little strange since GR cannot localize energy, and it explains why Einstein performed instead a classical calculation of a spinning rod losing energy by the emission of GWs that made use of the Poynting vector, which we discussed in §5.5.2. This explicitly involves the existence of an auxiliary field, $\vec{h}$, since Poynting's vector is:

$$\vec{S} = \frac{v}{4\pi}(\vec{g} \wedge \vec{h}).$$

The facts that "plane waves are a very special, and artificial case of gravitational wave propagation," lead Eddington to consider *divergent* waves. And

> although the equations of the theory are the same as those occurring in
> the propagation of sound waves, there is no propagation of gravitational
> waves uniformly in all directions like a spherical sound wave.

---

[72]M Ibison, H E Puthoff & S R Little, "The speed of gravity revisited."
[73]S Carlip, "Aberration and the speed of gravity," arXiv:gr-qc/9909087v2.
[74]A S Eddington, "The propagation of gravitational waves," *Proc Roy Soc London* Series A (1922) 268-282.
[75]H Weyl, *Space Time Matter* (Dover, New York, 1952) p 252.

However, we should not expect sound waves to be uniformally propagated in all directions since they are *longitudinal* with material motion in the direction of propagation.

To consider this possibility further, suppose that the gravitational field is the negative gradient of the *velocity* potential, ϕ, i.e.,

$$\vec{g} = -\nabla\phi.$$

Let $P = \dot{\phi}$ be the power loss due to the emission of GWs, where the rate of power loss is given by:

$$\dot{P} + v^2 \nabla \cdot \vec{g} = 0.$$

Replacing P and $\vec{g}$ by their definitions in terms of the velocity potential, ϕ, leads immediately to the scalar wave equation,

$$\ddot{\phi} - v^2 \nabla^2 \phi = 0.$$

These are longitudinal waves that propagate at a finite speed v, whatever that may turn out to be. The bad news is that they are non-polarized, but this is more than compensated by the fact that you do not have to explain gravitational aberration, and have the possibility of impacting matter just like pressure waves. Albeit, this is skirting entirely the basic question of what is causing the polarizing of the GWs.

Each model has its own attributes, and short-comings. The analogy with EM waves necessarily attribute to GWs the phenomena of diffraction and aberration. EM waves can be shielded whereas GWs cannot. Rather, if GWs are mechanical transverse waves, they need a material medium, and propagate by means of vibrations of the medium normal to the direction of propagation. The type of polarization would then be characteristic of the material medium, whereas for EM waves, the polarizations are given by the Grassmann force, and are equal and opposite in directions normal to the motion.

Yet, none of these models would explain the lack of shielding of gravity, and what we are left with is a compromise: The optical properties of gravity.

Index

www.ingramcontent.com/pod-product-compliance
Lightning Source LLC
Chambersburg PA
CBHW060822170526
45158CB00001B/55